Supermanifolds
Theory and Applications

Supermanifolds
Theory and Applications

Alice Rogers
King's College London

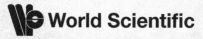
World Scientific

NEW JERSEY · LONDON · SINGAPORE · BEIJING · SHANGHAI · HONG KONG · TAIPEI · CHENNAI

Published by

World Scientific Publishing Co. Pte. Ltd.

5 Toh Tuck Link, Singapore 596224

USA office: 27 Warren Street, Suite 401-402, Hackensack, NJ 07601

UK office: 57 Shelton Street, Covent Garden, London WC2H 9HE

British Library Cataloguing-in-Publication Data
A catalogue record for this book is available from the British Library.

ISBN-13 978-981-02-1228-5
ISBN-10 981-02-1228-3

Printed in Singapore.

to Richard

Preface

The subject of supermanifolds has to some extent 'just growed' like Topsy, as the idea of adjoining anticommuting variables to conventional commuting variables proved useful in a variety of contexts. This has led to a bewildering variety of approaches which has often obscured the underlying unity of the ideas. Supermanifolds have been with me through much of this period of growth, and now in writing this book I intend not a formal mathematical treatise, but rather a working man or woman's guide to the geometry and analysis of supermanifolds, together with applications of the theory. Supermanifolds continue to find new uses; the underlying ideas have proved robust, powerful and adaptable. My aim is to provide a unified picture, distilling the key ideas from a welter of sources. I have tried to give references to relevant original work, and can only apologise for any failures, which will have been unintentional.

I am very grateful for many conversations about supermanifolds that I have enjoyed and benefited from over the years – these have been with almost everyone whose work is referred to in this book, a sad exception being Berezin. I would particularly like to thank Bryce DeWitt and Marjorie Batchelor, both of whom have patiently explained many things to me. I am also grateful to my colleagues in the mathematics department of King's College London for a stimulating and agreeable working environment.

F. A. Rogers
King's College London
2006

Contents

Chapter 1

Introduction

This book contains an account of the notions involved in constructing a theory of supermanifolds and the associated machinery and techniques of differential geometry, together with applications to various areas of physics, including supersymmetry and the quantization of systems with symmetry, and to classical geometry.

The concept of supermanifold involves an extension of a classical manifold to include some notion of anticommuting coordinate; indeed more generally the prefix 'super' is used with many mathematical objects to denote an extension from commutativity to graded commutativity, or to a controlled mixture of both commutativity and anticommutativity. The study of supermanifolds involves mathematical ideas from geometry, analysis, algebra and topology. While much of the original motivation came from particle physics, the concepts and language of supermanifolds have proved powerful in many parts of theoretical physics and pure mathematics, and the range of influence continues to broaden.

Historically anticommuting variables, and some of the constructions now distinguished by the prefix 'super', appeared in mathematics many years before the development of supersymmetry in physics triggered an explosion of interest in super mathematics. Of course anticommuting objects appear in many areas of geometry and algebra, examples include differential forms, the use of the exterior algebra over a Lie algebra in Lie algebra cohomology and the Weil model of equivariant cohomology. But perhaps the earliest step in 'super' mathematics was Cartan's recognition that a Clifford algebra could be represented on a Grassmann algebra if one included a notion of differentiation with respect to a generator as well as multiplication [29], an idea that was to reappear decades later in connection with fermion anticommutation relations. In his seminal work on quantum fields Schwinger

[139] introduced anticommuting variables in order to extend to fermions his treatment of quantum fields using Green's functions and sources. Differential calculus for functions of anticommuting variables was introduced by Martin [102] who extended Feynman's path-integral method of quantization to systems containing fermions and thus needed a 'classical' fermion to quantize. Anticommuting variables were used by a number of other authors to develop fermionic quantization in close analogy to methods for bosonic quantization using conventional, commuting variables; an extensive and pioneering study was made by Berezin [16].

A supersymmetric theory enjoys invariance under a symmetry which exchanges bosonic and fermionic degrees of freedom; as a result, an approach which treats fermions and bosons on an equal footing (as is the case when commuting and anticommuting variables are used) is likely to be particularly useful, and it has indeed been the case that super mathematical ideas have proved effective in the study of supersymmetry. Where geometrical ideas are involved in a supersymmetric model the anticommuting extension must respect and possibly develop this geometry, and thus what have become known as supermanifolds are required, together with much of the machinery of differential geometry. As interest in supersymmetric models took off in the physics community following the appearance of the pioneering models in the early 1970's [152, 158], there was a correspondingly rapid development of super geometry and other areas of super mathematics. The importance of anticommuting variables in supersymmetry can also be seen quite independently of any specific treatment of fermions, by considering the nature of the group of symmetries involved; at the infinitesimal level these form a super Lie algebra, that is, an algebra whose generators can be classified as either odd or even, and which closes under commutation of even generators and anticommutation of odd generators; the natural way to regard a group made from such generators is to associate commuting and anticommuting parameters respectively with the even and odd generators, leading to the concept of super Lie group. Although it is possible to handle supersymmetry without using anticommuting variables, their use often suggests by analogy some new approach to be tried, and has been a fruitful source of both conceptual and technical ideas.

At its simplest super mathematics extends classical ideas to a Z_2 graded setting, introducing a notion of even and odd, together with a rule that an extra sign factor appears whenever two odd elements are interchanged. Some of the development proceeds by a straightforward analogy with the classical, purely commuting case, with little more required than the insertion of the correct sign factors; however one aim of this book is to make it clear that the interesting and powerful parts of super mathematics are those where a straightforward analogy with classical mathematics is not possible, or does not give a full picture. There are various characteristic features of super mathematics which particularly stand out, which will occur repeatedly in the course of the book. These include the notion of super derivative (which provides a square root of a conventional derivative and also allows a representation of canonical anticommutation relations and Clifford algebras) and the Berezin integral which preserves certain characteristics of classical integration, but also has unexpected but valuable features. There is also the concept of supertrace, which leads to cancellations between odd and even sectors. These features, which are interrelated, are the key ingredients of many application of super mathematics both in geometry and in theoretical physics.

In super geometry there are two rather different, but essentially equivalent, approaches to supermanifold. In the first, which will be referred to as the *concrete* approach, a supermanifold is a set, more specifically it is a manifold modelled on some flat 'superspace' so that it has local coordinates some of which take values in the even and some in the odd part of a Grassmann algebra. In the second approach to supermanifolds, which will be referred to as the *algebro-geometric* approach, it is the sheaf of functions on a manifold which is extended, rather than the manifold itself. Here super geometry is distinguished from more general non-commutative geometry in which only an algebro-geometric approach seems to exist, with the non-commutativity expressed in terms of rings of 'functions'.

In this book both approaches to supermanifold are described. Large swathes of the subject are independent of the approach, but the emphasis in this book is on the concrete approach, because of the nature of the applications considered. Each approach has its protagonists, but in general a multicultural point of view, using the language of whichever of the two approaches best suits the matter in hand, is possible because there is a precise correspondence between algebro-geometric supermanifolds and concrete supermanifolds, as is explained in Chapter 8. Given this choice of approach, which does not exist for more general non-commutative geometry, it seems sensible to exploit all possibilities. The algebro-geometric approach has greater mathematical elegance and simplicity, because in its simplest form there is no need to introduce an auxiliary Grassmann algebra. However certain useful concepts, such as a point in a supermanifold, or an odd constant, are more complicated to define. To an accomplished algebraic geometer this

will not not be a problem, but to many who might use supermanifolds this
adds an unnecessary complication when a more direct approach is possible.
Moreover, in may physical applications it is in fact necessary to introduce
an auxiliary Grassmann algebra (or extra odd dimensions) when using the
algebro-geometric approach, and so the purity of the approach is diluted.
The concrete approach, in which a supermanifold is a set, and a super
Lie group is a group, has a psychological advantage in some contexts in
suggesting analogies with steps taken in classical differential geometry. It
also allows rather simpler terminology, for instance when using functions
between supermanifolds, and so makes it easier to give a direct description
of various applications and techniques. But it must be emphasised that
this is a question of choice of language, not an intrinsic difference. As an
example, consider the $(1, 1)$-dimensional super group of super translations.
Anticipating some notation and terminology, this is readily described in the
concrete approach as the set $\mathbb{R}_S^{1,1}$ with group action

$$(x; \xi) \circ (y; \eta) = (x + y + \xi\eta; \xi + \eta),$$

while in the algebro-geometric approach the same object is captured in a
less direct way. The main objection to the concrete approach is that it
carries the extra baggage of a Grassmann algebra whose individual elements do not individually signify as much as they might appear to; the
myriad coefficients, real or complex, of the Grassmann algebra expanded
with respect to some basis, do not carry useful information. While this
is true, broadly speaking the notion of concrete supermanifold is independent of the choice of Grassmann algebra since a particular topology (due
to DeWitt [43]), which does not distinguish between the various nilpotent
elements of the Grassmann algebra, is used. One distils out the mean-
inglessness of the Grassmann detail by showing (following Batchelor [12])
that there is a natural sequence of supermanifolds modelled on Grassmann
algebras with increasing numbers of generators, and taking the inductive
limit.

As remarked above, there are certain areas of supermanifold theory
where an auxiliary Grassmann algebra (or equivalent) is required in both
approaches. While in the smooth setting a theorem due to Batchelor [11]
shows that the data of a supermanifold is simply that of a vector bundle
over a classical manifold, in the complex setting one immediately encounters
supermanifolds whose data includes anticommuting parameters or moduli.
To handle such supermanifolds in the algebro-geometric approach requires
either the introduction of an auxiliary Grassmann algebra or the considera-

tion of families rather than individual supermanifolds, in which case auxiliary coordinates (even, odd or both) are required. In physical applications, and also in applications to classical geometry, the principal mechanism for extracting a real or complex number from the nilpotent superfluity is the Berezin integral. This a formal algebraic integral, not the limit of a sum depending on the details of any Grassmann algebra element. It is closely related to the notion of supertrace, as is explained in Section 11.1.

The use of supergeometric language, concepts and constructions is in a sense optional; the effect of adding odd dimensions and so on can always, or at least almost always, be produced by an alterative mathematical construction, and one does not really expect that anticommuting variables will model reality in the same way that real variables seem to. A strong argument for taking the 'super' point of view is that it suggests new approaches by analogy with procedures in classical mathematics. In a number of important situations super methods have proved very powerful. These include applications to classical geometry, supersymmetry and and the treatment of systems with gauge symmetry using ghosts and BRST cohomology. In this context it is interesting to quote Voronov's comment [153] in connection with the Faddeev-Popov quantization of gauge fields [49] that 'although he obtained similar results, DeWitt was unable to present them in a convenient form because he was unaware of the concept of the Berezin integral; see the author's forward to the Russian translation [42]' of [41].

Although a number of books and articles on supermanifolds have appeared, most are devoted to one particular approach, with the majority taking the algebro-geometric approach. In this book, while mostly using the language of the concrete approach, the aim is to take stock of a wide range of ideas and applications in supermanifold theory, hopefully throwing light on a concept that over the years has proved robust and adaptable to a number of purposes and abstracting the essential ideas from the various slightly different approaches. In general the style adopted has not assumed great mathematical sophistication on the part of the reader; this may irritate the accomplished mathematician, but is intended to increase the accessibility of the material for those who might use it in a variety of contexts. Although this book aims to provide a reasonably comprehensive survey of supermanifolds and their uses, examples considered are selected for their interest or their usefulness and there is no attempt to provide a complete taxonomy.

While detailed references will of course be given as the book proceeds, it seems appropriate to mention in the introduction some of the major landmarks in the history of the subject, as well as the principal books. Supermanifolds were introduced into the mathematical literature by Berezin and Leĭtes [19] and by Kostant [95], who used the term graded manifold. Both these authors used the algebro-geometric approach. Inspired by physics, DeWitt developed the theory of supermanifolds in the concrete approach, introducing the crucial topology which underpins the relation between the algebro-geometric and concrete approach [43]; Rogers developed the concrete approach from a slightly different point of view, imitating as closely as possible the construction of a classical manifold [118]. Berezin's extensive work on super mathematics was collected together after his untimely death in 1980 [17]. An early review article is that of Leĭtes [99]. More recent books on supermanifolds include that of Bartocci, Bruzzo and Hernández Ruipérez [8] and that of Tuynman [149]. Supermanifolds in the algebro-geometric approach are a central topic in the book of Manin [100]. A treatment of supermanifolds may be found in a section by Deligne and Morgan (following Bernstein) in 'Quantum fields and strings: a course for mathematicians' [39]. Two key papers by Batchelor [11, 12] provide the key structure theorem for smooth supermanifolds and establish the relationship between the concrete and the algebro-geometric approach.

Many of the basic features of supermanifold theory are well understood, although there are areas, such as a systematic treatment of subsupermanifolds, which remain incomplete. The structures developed have proved useful in may contexts; one result of supermanifold theory is simply to confirm that a heuristic approach, using super ideas developed by simple analogy with classical ones, is generally valid and often, when combined with the features of super geometry which have no classical analogue, very powerful. In many cases it is not the fully general theory of supermanifolds which has proved of interest, rather it has been supermanifolds restricted by some particular condition, for instance the super covariant condition used to define a super Riemann surface or the dimensional restriction necessary for an odd symplectic supermanifold, which have proved fruitful. It seems likely that further developments of supermanifold theory will involve new structures of this nature.

Chapter 2

Super algebras

The study of superspace and supermanifolds involves from the outset replacing ordinary real or complex variables with elements of some algebra whose elements commute or anticommute among themselves. To make this idea of mixed commutative and anticommutative properties more precise it is useful to introduce the concept of a super commutative algebra, which is an algebra which splits into two parts, called even and odd, with odd elements anticommuting with one another and even elements commuting with all elements. In the concrete approach supermanifolds are locally modelled on a superspace which is built from the even and odd parts of a super commutative algebra, while in both approaches to supermanifolds the function sheaves on which so much else depends are sheaves of super commutative algebras. (In the mathematical literature the term $\mathbb{Z}_2$-graded may sometimes be used instead of super.)

This chapter introduces the concept of super algebra. Various further algebraic constructions, such as homomorphisms, derivations and modules of super algebras are described. The concept of super Lie algebra is defined, and the appropriate generalisation of linear algebra, including the theory of super matrices, is developed.

In general the convention will be used that there is an implied summation over repeated indices. Since there are many different kinds of indices, with varying ranges, explicit sums will be used when needed for clarity.

2.1 The definition of a super algebra

The starting point in the theory of super algebras is the concept of super vector space, which is a vector space which splits as a direct sum as in the following definition.

Definition 2.1.1   A *super vector space* $\mathbb{V}$ is a vector space together with a choice of two subspaces $\mathbb{V}_0$ and $\mathbb{V}_1$ of $\mathbb{V}$ such that

$$\mathbb{V} = \mathbb{V}_0 \oplus \mathbb{V}_1. \tag{2.1}$$

Elements of the subspace $\mathbb{V}_0$ are said to be *even* and elements of $\mathbb{V}_1$ are said to be *odd*. Also, an element V which belongs to $\mathbb{V}_i$ where $i = 0$ or 1 is said to be *homogeneous* and $|V|$, the $\mathbb{Z}_2$ *degree* of V, is defined to be equal to i. (Thus even elements have degree 0 and odd elements have degree 1.)

An algebra is a vector space whose elements can be multiplied; a super algebra, defined formally in the following definition, is a super vector space whose elements can be multiplied, with the degree of the product determined by the degree of the factors.

Definition 2.1.2   Suppose that $\mathbb{A}$ is an algebra over the real numbers $\mathbb{R}$ or the complex numbers $\mathbb{C}$. Then $\mathbb{A}$ is said to be a *super algebra* if it is also a super vector space (over the same field) and

$$\begin{aligned}
\mathbb{A}_0\mathbb{A}_0 \subset \mathbb{A}_0, && \mathbb{A}_0\mathbb{A}_1 \subset \mathbb{A}_1, \\
\mathbb{A}_1\mathbb{A}_0 \subset \mathbb{A}_1, && \mathbb{A}_1\mathbb{A}_1 \subset \mathbb{A}_0.
\end{aligned} \tag{2.2}$$

The super algebra $\mathbb{A}$ is said to be *super commutative* if, whenever A and B are homogeneous elements of $\mathbb{A}$,

$$AB = (-1)^{|A||B|}BA. \tag{2.3}$$

(Thus even elements commute with all elements, while odd elements anti-commute with one another; in particular the square of an odd element is always zero.) Relations such as (2.3) which depend on degree are extended to non-homogeneous elements by linearity. The degree $|A|$ of an element A is an element of Z_2, and any addition or multiplication of degrees is carried out modulo 2. Thus the defining conditions (2.2) of a super algebra can be summarised as

$$\mathbb{A}_i\mathbb{A}_j \subset \mathbb{A}_{i+j} \qquad i, j = 0, 1.$$

Examples of super commutative algebras include the exterior algebra $\Lambda(V)$ over a finite-dimensional vector space, and the algebra of differential forms on a manifold (under the exterior product). In both these cases the $\mathbb{Z}_2$ degree is simply the $\mathbb{Z}$ degree taken modulo two.

2.2 Homomorphisms and modules of super algebras

In this section various classes of maps between super vector spaces and super algebras are considered. Also, the notion of a super module over a super algebra is introduced, together with the appropriate class of mappings between such spaces. The starting point is the degree assigned to linear maps between super vector spaces.

Definition 2.2.1 Suppose that $\mathbb{V}, \mathbb{W}$ are super vector spaces and that f is a linear mapping of $\mathbb{V}$ into $\mathbb{W}$. Then f is said to be a *super vector space homomorphism*. If additionally f satisfies

$$|f(V)| = |V| \mod 2 \tag{2.4}$$

for all V in the super vector space $\mathbb{V}$, then f is said to be an *even* super vector space homomorphism. Similarly, if f satisfies

$$|f(V)| = 1 + |V| \mod 2 \tag{2.5}$$

for every V in $\mathbb{V}$, f is said to be an *odd* super vector space homomorphism.

If f is a super vector space homomorphism, then the degree of f is denoted $|f|$ and defined to be 0 if f is even and 1 if f is odd. Thus in general a super vector space homomorphism f satisfies

$$|f(V)| = |f| + |V| \tag{2.6}$$

where, as always with Z_2 degree, the addition is modulo 2. The proof of the following proposition is left as an exercise.

Proposition 2.2.2 *If f is a super vector space homomorphism then there exists a unique even vector space homomorphism f_0 and a unique odd vector space homomorphism f_1 such that $f = f_0 + f_1$.*

As usual a homomorphism of a space into itself is called an endomorphism, a bijective homomorphism whose inverse is also a homomorphism is called an isomorphism and an isomorphism which is also an endomorphism is called an automorphism. A particular class of super vector space endomorphisms of a super algebra is the class of super derivations. Such endomorphisms are important when considering vector fields on supermanifolds.

Definition 2.2.3 Let $\mathbb{A}$ be a super commutative algebra; then a mapping $f : \mathbb{A} \to \mathbb{A}$ is said to be a *super derivation* if it is a super vector space homomorphism and additionally it obeys the super Leibniz condition

$$f(A_1 A_2) = f(A_1)A_2 + (-1)^{|A_1||f|}A_1 f(A_2) \tag{2.7}$$

for all $A_1, A_2 \in \mathbb{A}$.

A super derivation is a vector space homomorphism, but not a super algebra homomorphism. These, as the following definition shows, are mappings which preserve the product.

Definition 2.2.4　　Let $\mathbb{A}$ and $\mathbb{B}$ be super algebras. Then a mapping $f : \mathbb{A} \to \mathbb{B}$ of definite parity is said to be a *super algebra homomorphism* if it is a super vector space homomorphism and additionally, for all $A_1, A_2 \in \mathbb{A}$,

$$f(A_1 A_2) = (-1)^{|f||A_1|}f(A_1)f(A_2). \tag{2.8}$$

A module is a standard concept in algebra. In general a module resembles a vector space, except that the scalars are now elements of an algebra rather than of a field. Because the algebra will generally have non-invertible elements modules do not always have bases or well-defined dimension. In the super case, there is an extra requirement of compatible degree.

Definition 2.2.5　　(a) Suppose that $\mathbb{V}$ is a super vector space and that $\mathbb{A}$ is a super commutative algebra. Then $\mathbb{V}$ is said to be a *left super $\mathbb{A}$-module* if there exists a mapping

$$\mathbb{A} \times \mathbb{V} \to \mathbb{V}$$
$$(A, V) \mapsto AV \tag{2.9}$$

such that, for all A_1, A_2 in $\mathbb{A}$ and all V in $\mathbb{V}$

$$|A_1 V| = |A_1| + |V|$$
$$\text{and} \quad A_1(A_2 V) = (A_1 A_2)V. \tag{2.10}$$

(b) Suppose there exist r elements $C_1, \dots, C_r$ of $\mathbb{V}_0$ and s elements $C_{r+1}, \dots, C_{r+s}$ of $\mathbb{V}_1$ such that each element V of $\mathbb{V}$ can be expressed as

$$V = \sum_i^{r+s} V^i C_i \tag{2.11}$$

for a unique element $(V^1, \dots, V^{r+s})$ of $\mathbb{A}^{r+s}$. Then $\mathbb{V}$ is said to be a *free super $\mathbb{A}$-module of dimension* (r, s), and the set $\{C_1, \dots, C_{r+s}\}$ is said to be a (r, s) *super basis*.

With respect to a particular super basis, an element V of $\mathbb{V}$ may be expressed as the $(r + s)$-tuple $(V^1, \ldots, V^{r+s})$. If V is even then $(V^1, \ldots, V^{r+s})$ is an element of the 'superspace'

$$\mathbb{A}^{r,s} = \underbrace{\mathbb{A}_0 \times \cdots \times \mathbb{A}_0}_{r \text{ copies}} \times \underbrace{\mathbb{A}_1 \times \cdots \times \mathbb{A}_1}_{s \text{ copies}} . \tag{2.12}$$

Free super modules appear frequently in supermanifold theory. For instance the tangent space at a point in a supermanifold is a free super module of the same dimension as the supermanifold.

For a non-commutative algebra (such as $\mathbb{A}$) one must distinguish between left and right $\mathbb{A}$-modules. The modules of Definition 2.2.5 are left $\mathbb{A}$-modules; however, because of the super-commutativity of $\mathbb{A}$, such a module can also be given the structure of a right module by defining the map

$$\mathbb{V} \times \mathbb{A} \to \mathbb{V}$$
$$(V, A) \mapsto (-1)^{|V||A|} AV . \tag{2.13}$$

(The super module property follows because

$$|VA| = |V| + |A|$$
$$\text{and} \quad V(A_1 A_2) = (V A_1) A_2 \tag{2.14}$$

for all V in $\mathbb{V}$ and A, A_1, A_2 in $\mathbb{A}$.) Thus, while an ordering convention is necessary, it will often not be specified whether a module is on the right or the left.

An important example of a super $\mathbb{A}$-module is the set of super derivations of $\mathbb{A}$ (Definition 2.2.3). The following proposition, which may readily be proved, shows that the set of super endomorphisms of a super algebra $\mathbb{A}$ has a natural $\mathbb{A}$-module structure. (It will be seen in Chapter 6 that these ideas allow a natural extension of the usual notion of vector field to supermanifolds, with $\mathbb{A}$ the algebra of supersmooth functions.)

Proposition 2.2.6 *The set* $\mathrm{Der}(\mathbb{A})$ *of super derivations of* $\mathbb{A}$ *is a left super* $\mathbb{A}$*-module with*

$$(AP)B = AP(B) \tag{2.15}$$

for all A, B *in* $\mathbb{A}$ *and* P *in* $\mathrm{Der}(\mathbb{A})$.

Both the exterior derivative of differential forms on a manifold and the interior derivative along a vector field are examples of odd derivations of the super algebra of differential forms.

To complete this section the definition of a homomorphism of super $\mathbb{A}$-modules will now be given. Such mappings are referred to as super $\mathbb{A}$-linear mappings.

Definition 2.2.7 Suppose that $\mathbb{A}$ is a super commutative algebra and that $\mathbb{V}$ and $\mathbb{W}$ are super $\mathbb{A}$ modules. Then a mapping $f : \mathbb{V} \to \mathbb{W}$ is said to be a *super $\mathbb{A}$-linear mapping* of super $\mathbb{A}$-modules if f is a super vector space homomorphism and also

$$f(AV) = (-1)^{|A||f|} A f(V) \qquad (2.16)$$

for all A in $\mathbb{A}$ and all V in $\mathbb{V}$.

The super matrices of the following section correspond to such transformations referred to specific bases in the usual way.

2.3 Super matrices

If $\mathbb{A}$ is a super commutative algebra, super matrices are matrices with entries in $\mathbb{A}$ which define even homomorphisms of free super $\mathbb{A}$-modules in terms of particular bases. In most applications to supermanifolds the super commutative algebra in question will be a Grassmann algebra.

Definition 2.3.1 A $(p,q) \times (r,s)$ *super matrix* over a super commutative algebra $\mathbb{A}$ is a $(p+q) \times (r+s)$ matrix $\mathcal{M}$ whose entries are elements of $\mathbb{A}$ and which has the block-diagonal form

$$\begin{pmatrix} \mathcal{M}_{0,0} & \mathcal{M}_{0,1} \\ \mathcal{M}_{1,0} & \mathcal{M}_{1,1} \end{pmatrix} \qquad (2.17)$$

with $\mathcal{M}_{0,0}$ a $p \times r$ matrix with even entries, $\mathcal{M}_{0,1}$ a $p \times s$ matrix with odd entries, $\mathcal{M}_{1,0}$ a $q \times r$ matrix with odd entries and $\mathcal{M}_{1,1}$ a $q \times s$ matrix with even entries. Such a super matrix is said to have order $(p,q) \times (r,s)$.

Suppose that $\mathbb{V}$ and $\mathbb{W}$ are free super $\mathbb{A}$-modules of dimension (p,q) and (r,s) respectively with super bases $\{C_1, \ldots, C_{p+q}\}$ and $\{B_1, \ldots, B_{r+s}\}$. Then an even super $\mathbb{A}$-linear mapping $f : \mathbb{V} \to \mathbb{W}$ can be expressed by the $(p,q) \times (r,s)$ super matrix $\left(f_i{}^j\right)$ where

$$f(C_i) = \sum_{j=1}^{r+s} f_i{}^j B_j \, , \, i = 1, \ldots p+q \, . \qquad (2.18)$$

Addition and multiplication of super matrices is defined in a way analogous to that for conventional matrices, with matching of super order being required so that the resulting matrices are super matrices and not simply matrices with $\mathbb{A}$-valued entries. Thus the sum of two super matrices of the same order will be a super matrix of that order, while the product of a $(p,q) \times (r,s)$ super matrix $\mathcal{M}$ with a $(r,s) \times (t,u)$ super matrix $\mathcal{N}$ will be a $(p,q) \times (t,u)$ super matrix $\mathcal{M}\mathcal{N}$. The product of super matrices corresponds to the combination of $\mathbb{A}$-linear transformations and so has the usual algebraic properties.

In supermanifold theory the most important super matrices are square super matrices; the set $\mathcal{M}((p,q),\mathbb{A})$ of $(p,q) \times (p,q)$ super matrices over $\mathbb{A}$ is closed under multiplication, and the subset $\mathrm{GL}(p,q;\mathbb{A})$ of invertible matrices is a group with identity element the matrix with the unit element of $\mathbb{A}$ along the leading diagonal and zeros elsewhere (denoted $I_{p,q}$ or simply I). When $\mathbb{A}$ is of suitable form, $\mathrm{GL}(p,q;\mathbb{A})$ will be a super Lie group of the kind discussed in Chapter 9.

Further properties of super matrices which depend on $\mathbb{A}$ being some Grassmann algebra are considered in Chapter 3.

2.4 Super Lie algebras and super Lie modules

In this final section on the basic algebraic structures involved in supermanifold theory various super extensions of the concept of Lie algebra are introduced. The simplest of these is the notion of super Lie algebra, which is a super vector space on which a super anticommutative Lie bracket is defined which satisfies a super Jacobi identity.

Definition 2.4.1   A *super Lie algebra* is a super vector space $\mathbb{U}$ together with a binary operation

$$\mathbb{U} \times \mathbb{U} \to \mathbb{U}$$
$$(X,Y) \mapsto [X,Y] \tag{2.19}$$

such that for all X, Y in $\mathbb{U}$

$$|[X,Y]| = |X| + |Y|$$
$$\text{and} \quad [X,Y] = -(-1)^{|X||Y|}[Y,X], \tag{2.20}$$

and for all X, Y, Z in $\mathbb{U}$ the super Jacobi identity

$$(-1)^{|X||Z|} [X, [Y, Z]] + (-1)^{|Z||Y|} [Z, [X, Y]] + (-1)^{|Y||X|} [Y, [Z, X]] = 0 \tag{2.21}$$

is satisfied.

This definition has the standard structure of a super version of a standard mathematical object; that is, every interchange of two elements is accompanied by the appropriate sign factor, as may be seen by multiplying through by a factor $(-1)^{|X||Z|}$. A more fundamental reason comes from considering super commutators.

There is an extensive literature on super Lie algebras (which are also known in the literature as Lie super algebras). The classification of finite-dimensional simple super Lie algebras was discovered by Kac [86]; a detailed account of this work may be found in the book of Scheunert [136]. This book also describes aspects of the representation theory of super Lie algebras. An extensive discussion of super Poincaré algebras is given in Cornwell's book [34], which also considers semisimple super Lie algebras.

The representation theory of super Lie algebras is based on the important result (whose proof is left to the reader) that the set of super endomorphisms of a super vector space form a super Lie algebra.

Proposition 2.4.2 *Let $\mathbb{V}$ be a (p, q)-dimensional super vector space. Then the set $\mathfrak{g}$ of super endomorphisms of $\mathbb{V}$ has the structure of a super Lie algebra of dimension $(p^2 + q^2, 2pq)$.*

As one might expect, super Lie algebras play an important rôle in the theory of super Lie groups; for some classes of super Lie groups (H^∞ Lie groups in the terminology of Chapter 9) super Lie algebras are sufficient. However for more general super Lie groups, and for some aspects of the analysis on all super Lie groups, it is useful to introduce the concept of super Lie module. Such a module is both a super Lie algebra and a module over some super algebra, with the two structures intertwined.

Definition 2.4.3 Suppose that $\mathbb{A}$ is a super algebra and that $\mathfrak{u}$ is a super Lie algebra which is also a super $\mathbb{A}$ module such that

$$[AU_1, U_2] = A[U_1, U_2] \tag{2.22}$$

for all A in $\mathbb{A}$ and U_1, U_2 in $\mathfrak{u}$. Then $\mathfrak{u}$ is said to be a *super Lie module* over $\mathbb{A}$.

Super Lie modules are important in supermanifold theory because the vector fields on a supermanifold form a super Lie module which is locally free, as is shown in Chapter 6. The following proposition, which establishes that the super derivations of a super commutative algebra form a super Lie module, is useful in the study of vector fields.

Proposition 2.4.4 *Let* $\mathrm{Der}(\mathbb{A})$ *denote the set of super derivations of the super commutative algebra* $\mathbb{A}$. *Then* $\mathrm{Der}(\mathbb{A})$ *is a super Lie module with bracket operation defined by the super commutator*

$$[X, Y] = XY - (-1)^{|X||Y|}YX \qquad (2.23)$$

for X, Y *in* $\mathrm{Der}(\mathbb{A})$.

Proof It has already been shown in Proposition 2.2.6 that $\mathrm{Der}(\mathbb{A})$ is a super $\mathbb{A}$-module. Direct calculation shows that

$$[X, Y](AB) = ([X, Y]A)B + (-1)^{|A|(|X|+|Y|)}A([X, Y]B) \qquad (2.24)$$

for all A, B in $\mathbb{A}$, so that $[X, Y]$ is a derivation of $\mathbb{A}$ of degree $|X| + |Y|$. The graded antisymmetry

$$[X, Y] = -(-1)^{|X||Y|}[X, Y] \qquad (2.25)$$

follows immediately from the definition, while the graded Jacobi identity can be verified by direct calculation. ∎

Example 2.4.5 Suppose that $\mathfrak{u}$ is a super Lie algebra, and that $\mathbb{A}$ is a super algebra. Then $\mathbb{A} \otimes \mathfrak{u}$ is a super Lie module over $\mathbb{A}$, with bracket defined by

$$[AX, BY] = (-1)^{|X||B|}AB[X, Y] . \qquad (2.26)$$

Super Lie modules with this tensor product structure are referred to as *product super Lie modules*.

Super Lie modules are also important in the study of super Lie groups. In Chapter 9 it is shown that the infinitesimal structure of the group is captured by a super Lie module. The analogy with the classical theory of Lie groups is quite close.

Chapter 3

Superspace

The particular super commutative algebras used to build concrete super-manifolds are usually Grassmann algebras. In the literature on super-manifolds a variety of such algebras have been employed, some finite-dimensional, some infinite-dimensional, some with Banach algebra properties, others with much coarser topologies. These algebras are briefly described here, and their merits and demerits discussed. Most emphasis is put on the particular real Grassmann algebra $\mathbb{R}_S$ which is much used in this book; this is defined in Section 3.1, together with the corresponding (m,n)-dimensional superspace $\mathbb{R}_S^{m,n}$ and its DeWitt topology which is defined in Section 3.2. In Section 3.3 the corresponding complex Grassmann algebra $\mathbb{C}_S$ is described, while in the final section the theory of super matrices with Grassmann algebra entries is developed.

3.1   Real Grassmann algebras

In this section the Grassmann algebra over a finite dimensional real vector space is defined, and various infinite-dimensional Grassmann algebras discussed. In the concrete approach to supermanifolds it is generally easier to model the supermanifold on a superspace built from an infinite-dimensional Grassmann algebra. While it might seem simpler to use a finite-dimensional algebra, there are difficulties which can arise when considering functions of such algebras, as will emerge in Chapter 4. Alternatively, one may use a Grassmann algebra with a variable, but finite, number of generators, and take some limit as the number tends to infinity.

Definition 3.1.1   For each finite positive integer L, $\mathbb{R}_{S[L]}$ denotes the Grassmann algebra over $\mathbb{R}$ with L generators. That is, $\mathbb{R}_{S[L]}$ is the algebra over $\mathbb{R}$ with generators

$$1, \beta_{[1]}, \ldots, \beta_{[L]}$$

and relations

$$1\beta_{[i]} = \beta_{[i]} = \beta_{[i]}1 \qquad i = 1, \ldots, L$$
$$\beta_{[i]}\beta_{[j]} = -\beta_{[j]}\beta_{[i]} \qquad i, j = 1, \ldots, L.$$

$$(3.1)$$

This is not the most elegant or abstract definition of this algebra, but is the most useful form for the constructions to be made below. A typical element X of $\mathbb{R}_{S[L]}$ may thus be expressed as

$$X = \sum_{\underline{\lambda} \in M_L} X_{\underline{\lambda}} \beta_{[\underline{\lambda}]} \qquad (3.2)$$

where $\underline{\lambda}$ is a multi index $\underline{\lambda} = \lambda_1 \ldots \lambda_k$ with $1 \leq \lambda_1 < \cdots < \lambda_k \leq L$, M_L is the set of all such multi indices (including the empty index $\emptyset$), each $X_{\underline{\lambda}}$ ($\underline{\lambda} \in M_L$) is a real number and $\beta_{[\underline{\lambda}]} = \beta_{[\lambda_1]} \ldots \beta_{[\lambda_k]}$ (with $\beta_{[\emptyset]} = 1$).

The Grassmann algebra $\mathbb{R}_{S[L]}$ is given the structure of a super commutative algebra by setting $\mathbb{R}_{S[L]} = \mathbb{R}_{S[L\,0]} \oplus \mathbb{R}_{S[L\,1]}$ with $\mathbb{R}_{S[L\,0]}$ consisting of sums of combinations of even numbers of anticommuting generators, and $\mathbb{R}_{S[L\,1]}$ of sums of combinations of odd numbers of anticommuting generators. That is, if $M_{L,0}$ is the set of multi indices in M_L which contain an even number of indices while $M_{L,1}$ is the set of multi indices in M_L which contain an odd number of indices,

$$\mathbb{R}_{S[L\,0]} = \left\{ x \,\middle|\, x \in \mathbb{R}_{S[L]}, \; x = \sum_{\underline{\lambda} \in M_{L\,0}} x_{\underline{\lambda}} \beta_{[\underline{\lambda}]} \right\}$$

$$\mathbb{R}_{S[L\,1]} = \left\{ \xi \,\middle|\, \xi \in \mathbb{R}_{S[L]}, \; \xi = \sum_{\underline{\lambda} \in M_{L\,1}} \xi_{\underline{\lambda}} \beta_{[\underline{\lambda}]} \right\}. \qquad (3.3)$$

Here the convention is used that lower case Latin letters denote even variables and lower case Greek letters denote odd variables (while capital letters will denote elements of either parity, or of no definite parity). A complex Grassmann algebra $\mathbb{C}_L$ is similarly defined as the Grassmann algebra over $\mathbb{C}$ with L anticommuting generators, as will be seen in Section 3.3.

Given a super commutative algebra $\mathbb{A}$, the corresponding (m,n)-dimensional superspace can be defined to be the space

$$\mathbb{A}^{m,n} = \underbrace{\mathbb{A}_0 \times \cdots \times \mathbb{A}_0}_{m \text{ copies}} \times \underbrace{\mathbb{A}_1 \times \cdots \times \mathbb{A}_1}_{n \text{ copies}} \tag{3.4}$$

with m said to be the even dimension and n the odd dimension of the superspace. Using the specific super commutative algebra $\mathbb{R}_{S[L]}$, it now becomes natural to consider flat (m,n)-dimensional superspace to be the space

$$\mathbb{R}^{m,n}_{S[L]} = \underbrace{\mathbb{R}_{S[L\,0]} \times \cdots \times \mathbb{R}_{S[L\,0]}}_{m \text{ copies}} \times \underbrace{\mathbb{R}_{S[L\,1]} \times \cdots \times \mathbb{R}_{S[L\,1]}}_{n \text{ copies}}, \tag{3.5}$$

with a typical element denoted $(x^1, \ldots, x^m; \xi^1, \ldots, \xi^n)$, or more briefly $(x; \xi)$. At this stage m and n are both arbitrary non-negative integers. In Chapter 13 it will be apparent that particular values of m and n are often appropriate in supersymmetric physics; there may also be restrictions on dimension when further structures are required.

It is useful to observe that there is a unique algebra homomorphism ϵ of $\mathbb{R}_{S[L]}$ onto $\mathbb{R}$ which maps the identity element 1 onto 1 and all the generators $\beta_{[i]}, i = 1, \ldots, L$ to zero. Thus

$$\epsilon : \mathbb{R}_{S[L]} \to \mathbb{R}$$

$$\sum_{\underline{\lambda} \in M_L} X_{\underline{\lambda}} \beta_{[\underline{\lambda}]} \mapsto X_\emptyset. \tag{3.6}$$

This map is known in mathematics as the augmentation map, but here, following DeWitt [43], it will be referred to as the *body map*. The complementary map, which projects onto the nilpotent elements, is the *soul map*

$$s : \mathbb{R}_{S[L]} \to \mathbb{R}_{S[L]}$$

$$X \mapsto X - \epsilon(X)1, \tag{3.7}$$

so that $X = \epsilon(X)1 + s(X)$. It is also useful to define

$$\epsilon_{m,n} : \mathbb{R}^{m,n}_{S[L]} \to \mathbb{R}^m$$

$$(x^1, \ldots, x^m; \xi^1, \ldots, \xi^n) \mapsto (\epsilon(x^1), \ldots, \epsilon(x^n)). \tag{3.8}$$

Because the product of $L+1$ anticommuting generators of $\mathbb{R}_{S[L]}$ must necessarily contain a repeated factor and thus be zero, the following proposition is immediate:

Proposition 3.1.2   *If* $X \in \mathbb{R}_{S[L]}$,

$$(s(X))^{L+1} = 0. \tag{3.9}$$

Because the soul of an element is always nilpotent, an element of $\mathbb{R}_{S[L]}$ whose body is zero will not be invertible. Conversely, if the body of an even element x of $\mathbb{R}_{S[L]}$ is not zero, then it is invertible, with inverse

$$x^{-1} = \frac{1}{\epsilon(x)} \left(1 - \frac{s(x)}{\epsilon(x)} + \left(\frac{s(x)}{\epsilon(x)} \right)^2 - \dots \right). \tag{3.10}$$

As the following proposition shows, it is possible to equip the Grassmann algebra $\mathbb{R}_{S[L]}$ with a norm in such a way that it becomes a Banach algebra.

Proposition 3.1.3   *Suppose* X *is an element of* $\mathbb{R}_{S[L]}$, *with*

$$X = \sum_{\underline{\lambda} \in M_L} X_{\underline{\lambda}} \beta_{[\underline{\lambda}]}.$$

Then if $\|X\|$ *is defined by*

$$\|X\| = \sum_{\underline{\lambda} \in M_L} |X_{\underline{\lambda}}|, \tag{3.11}$$

$\| \ \|$ *defines a norm on* $\mathbb{R}_{S[L]}$. *Also* $\mathbb{R}_{S[L]}$ *with this norm is a Banach algebra, that is,* $\|1\| = 1$ *and, for all* X *and* Y *in* $\mathbb{R}_{S[L]}$,

$$\|XY\| \leq \|X\|\|Y\|. \tag{3.12}$$

Much of this theorem follows from the standard ℓ^1 norm on a finite dimensional vector space. The algebraic aspect follows by linearity from the facts that $\|\beta_{[\underline{\mu}]}\| = 1$ and $\beta_{[\underline{\nu}]}\beta_{[\underline{\mu}]}$ is either zero or $\pm\beta_{[\underline{\lambda}]}$ for some multi index $\underline{\lambda}$.

It is possible to slightly modify the definition of a Grassmann algebra to allow an infinite number of generators.

Definition 3.1.4 $\mathbb{R}_S$ *denotes the algebra over* $\mathbb{R}$ *with generators*

$$1, \beta_{[1]}, \dots$$

and relations

$$1\beta_{[i]} = \beta_{[i]} = \beta_{[i]}1 \qquad i = 1, 2, \ldots$$

$$\beta_{[i]}\beta_{[j]} = -\beta_{[j]}\beta_{[i]} \qquad i, j = 1, 2, \ldots . \tag{3.13}$$

A typical element of $\mathbb{R}_S$ is thus

$$B = \sum_{\underline{\lambda} \in M_\infty} b_{\underline{\lambda}}\beta_{[\underline{\lambda}]} \tag{3.14}$$

where each $\underline{\lambda}$ is a finite multi index $\underline{\lambda} = \lambda_1 \ldots \lambda_k$ with $1 \leq \lambda_1 < \cdots < \lambda_k$, M_∞ is the set of all such multi indices (including the empty index $\emptyset$), each $b_{\underline{\lambda}}$ ($\underline{\lambda} \in M_L$) is a real number and $\beta_{[\underline{\lambda}]} = \beta_{[\lambda_1]} \ldots \beta_{[\lambda_k]}$ (with $\beta_{[\emptyset]} = 1$). The number of indices in the multi-index $\underline{\lambda}$ is denoted $\ell(\underline{\lambda})$, and the term $b_{\underline{\lambda}}\beta_{[\underline{\lambda}]}$ is said to have length $\ell(\underline{\lambda})$.

The Grassmann algebra $\mathbb{R}_S$ is a super commutative algebra, with $\mathbb{Z}_2$ degree defined as for the finite-dimensional algebra $\mathbb{R}_{S[L]}$. Also the body and soul maps may be defined in an analogous manner.

Elements of $\mathbb{R}_S$ are formal infinite sums which are added and multiplied according to the usual rules of algebra. These sums and products are well-defined because each term in the expansion (3.2) in terms of finite products of generators is well-defined. Questions of convergence only arise when one considers the topology on $\mathbb{R}_S$. Various topologies are discussed in the following section, where it will be seen that the most commonly used topology – the DeWitt topology – involves no convergence conditions. It is the lack of invertibility of the anticommuting generators which makes such a topology useful, because it has the effect that terms containing a given generator in a factor cannot contribute to terms without that generator in the product. The corresponding situation with a Clifford algebra would be quite different.

3.2 The topology of superspace

There are a number of different topologies one can use on the superspace $\mathbb{R}_S^{m,n}$. The most important topology is that introduced by DeWitt [43]; despite the fact that it is a non-Hausdorff topology, it will emerge in the following chapter that specific algebraic features make it the appropriate topology to use in many aspects of supermanifold theory.

Definition 3.2.1   A subset U of $\mathbb{R}_S^{m,n}$ is said to be open in the *DeWitt topology* on $\mathbb{R}_S^{m,n}$ if and only if there exists an open subset V of $\mathbb{R}^m$ such that

$$U = \epsilon_{m,n}{}^{-1}(V). \tag{3.15}$$

This topology is by far the most useful topology on superspace, but there are other possibilities which have been considered in the literature.

When using a finite-dimensional Grassmann algebra $\mathbb{R}_{S[L]}$, the superspace $\mathbb{R}_{S[L]}^{m,n}$ is also a finite-dimensional vector space, and one can simply use the usual finite vector space topology, but the DeWitt topology is also available. On superspaces built from infinite-dimensional Grassmann algebras there are various possible topologies in addition to the DeWitt topology. One approach (first considered in [118], and developed considerably by Jadczyk and Pilch [84]) is to restrict the algebra $\mathbb{R}_S$ in such a way that it can be given the structure of a Banach algebra. In [118] this was done by considering the subspace $\mathbb{R}_{S[\infty]}$ consisting of elements with finite ℓ_1 norm; the following theorem justifies this approach.

Theorem 3.2.2   *Let $\mathbb{R}_{S[\infty]}$ denote the vector subspace of $\mathbb{R}_S$ consisting of elements $B = \sum_{\underline{\mu} \in M_\infty} X_{\underline{\mu}} \beta_{[\underline{\mu}]}$ such that $\|X\|_1 =_{def} \sum_{\underline{\mu} \in M_\infty} \left| X_{\underline{\mu}} \right| < \infty$. Then $\mathbb{R}_{S[\infty]}$ forms a subalgebra of $\mathbb{R}_S$, and $\mathbb{R}_{S[\infty]}$ with norm $\| \ \|_1$ is a Banach algebra.*

Proof   Suppose that X and Y are elements of $\mathbb{R}_{S[\infty]}$ with $X = \sum_{\underline{\mu} \in M_\infty} X_{\underline{\mu}} \beta_{[\underline{\mu}]}$ and $Y = \sum_{\underline{\mu} \in M_\infty} Y_{\underline{\mu}} \beta_{[\underline{\mu}]}$. Then

$$XY = \sum_{\underline{\mu} \in M_\infty} \sum_{\underline{\nu} \in M_\infty} X_{\underline{\mu}} Y_{\underline{\nu}} \beta_{[\underline{\mu}]} \beta_{[\underline{\nu}]}. \tag{3.16}$$

Now $\beta_{[\underline{\mu}]} \beta_{[\underline{\nu}]}$ is either zero or equal to a product of some of the Grassmann generators $\beta_{[1]}, \beta_{[2]}, \ldots$, so that

$$\|XY\|_1 \leq \sum_{\underline{\mu} \in M_\infty} \sum_{\underline{\nu} \in M_\infty} \left| X_{\underline{\mu}} Y_{\underline{\nu}} \right|$$

$$= \sum_{\underline{\mu} \in M_\infty} \sum_{\underline{\nu} \in M_\infty} \left| X_{\underline{\mu}} \right| \left| Y_{\underline{\nu}} \right|$$

$$= \|X\|_1 \|Y\|_1$$

$$< \infty. \tag{3.17}$$

This shows that the product XY also lies in $\mathbb{R}_{S[\infty]}$ and thus $\mathbb{R}_{S[\infty]}$ is a subalgebra of $\mathbb{R}_S$. Equation (3.17) also shows that the product of two elements in $\mathbb{R}_{S[\infty]}$ has the Banach algebra property

$$\|XY\|_1 \leq \|X\|_1 \|Y\|_1. \tag{3.18}$$

Direct calculation shows that $\|1\|_1 = 1$. Thus $\mathbb{R}_{S[\infty]}$ with norm $\| \ \|_1$ is a Banach algebra. ∎

Such Banach Grassmann algebras have been considered in great detail by Pestov [109].

A third possibility, developed by Inoue and Maeda [83] is to give $\mathbb{R}_S$ the structure of a Fréchet space; briefly, a Fréchet space is a vector space on which a family of seminorms is defined such that the seminorms of the distance between two distinct points are never all zero, and such that the corresponding natural topology is metrizable. Inoue and Maeda give $\mathbb{R}_S$ the structure of a Fréchet space by considering the finite-dimensional subspaces $\mathbb{R}_{S[j]}^f, j = 0, 1, \ldots$ consisting of terms whose length (in the expansion (3.14)) does not exceed j. The seminorms $\rho_j, j = 0, 1, \ldots$ are then defined by projection onto $\mathbb{R}_{S[j]}^f$ followed by taking the ℓ_1 norm.

A further approach to superspace is developed by Nagamachi and Kobayashi [107], who consider an inductive limit (as L tends to ∞) of the finite-dimensional algebras $\mathbb{R}_{S[L]}$. This approach is closely related to the approach of Batchelor [12] and Rogers [122] where inductive limits of supermanifolds modelled on $\mathbb{R}_{S[L]}$ are considered. Nagamachi and Kobayashi also consider algebras with more general commutativity properties.

Amid this confusing welter of possible Grassmann algebras on which to model supermanifolds, the infinite-dimensional Grassmann algebra $\mathbb{R}_S$ with the DeWitt topology seems to have the necessary analytic features without superfluous analytical encumbrances, and will be the one largely employed in this book.

3.3 Complex Grassmann algebras

The basic definition of complex Grassmann algebra closely resembles that of a real Grassmann algebra. For completeness it will now be given.

Definition 3.3.1 For each finite positive integer L, $\mathbb{C}_{S[L]}$ denotes the Grassmann algebra over $\mathbb{C}$ with L generators. That is, $\mathbb{C}_{S[L]}$ is the algebra over $\mathbb{C}$ with generators

$$1, \beta_{[1]}, \ldots, \beta_{[L]}$$

and relations

$$1\beta_{[i]} = \beta_{[i]} = \beta_{[i]}1 \qquad i = 1, \ldots, L$$
$$\beta_{[i]}\beta_{[j]} = -\beta_{[j]}\beta_{[i]} \qquad i, j = 1, \ldots, L. \tag{3.19}$$

A typical element X of $\mathbb{C}_{S[L]}$ may thus be expressed as

$$X = \sum_{\underline{\lambda} \in M_L} X_{\underline{\lambda}} \beta_{[\underline{\lambda}]} \tag{3.20}$$

where each coefficient $X_{\underline{\lambda}}$ is a complex number. The definition may readily be extended to give a definition of the complex Grassmann algebra $\mathbb{C}_S$ with an infinite set of generators in strict analogy to $\mathbb{R}_S$.

The notion of complex conjugation is important, and can be defined in a variety of ways. The approach taken here (which, up to sign conventions, is that usually used in physics) has several useful properties. An alternative definition, with interesting physical applications, is given by Kleppe and Wainwright in [91].

Definition 3.3.2   *Complex conjugation* is the unique $\mathbb{R}$-linear map of $\mathbb{C}_S$ into itself such that,

$$(ca)^* = c^* a^* \quad \text{for all } c \in \mathbb{C} \text{ and } a \in \mathbb{C}_S$$
$$1^* = 1$$
$$\beta_{[i]}^* = i\beta_{[i]} \text{ for each generator } \beta_{[i]}$$
$$\text{and } (ab)^* = b^* a^* \quad \text{for all } a, b \in \mathbb{C}_S, \tag{3.21}$$

where c^* denotes the complex conjugate of $c \in \mathbb{C}_S$.

Theorem 3.3.3   *conjugation is an involution, that is,*

$$c^{**} = c \text{ for all } c \in \mathbb{C}_S. \tag{3.22}$$

Proof It follows from (3.21) that

$$(a\beta_{[\underline{\lambda}]})^* = a^* i^{\ell(\underline{\lambda})} \beta_{[\overleftarrow{\underline{\lambda}}]}. \tag{3.23}$$

where $\beta_{[\overleftarrow{\underline{\lambda}}]} = \beta_{[\lambda_{\ell(\underline{\lambda})}]} \cdots \beta_{[\lambda_1]}$. Hence

$$(a\beta_{[\underline{\lambda}]})^{**} = a(-i)^{\ell(\underline{\lambda})} i^{\ell(\underline{\lambda})} \beta_{[\underline{\lambda}]} = a\beta_{[\underline{\lambda}]}. \tag{3.24}$$

∎

In order to identify the real Grassmann algebra $\mathbb{R}_S$ as a subalgebra of $\mathbb{C}_S$, an element C of $\mathbb{C}_S$ is defined to be *real* if and only if

$$C^* = i^{|C|}C. \tag{3.25}$$

With this definition, if C_1 and C_2 are real then the sum $C_1 + C_2$ is clearly real and also, since

$$(C_1 C_2)^* = C_2{}^* C_1{}^*$$
$$= i^{|C_1|} i^{|C_2|} C_2 C_1$$
$$= i^{(|C_1|+|C_2|) \bmod 2} C_1 C_2, \tag{3.26}$$

the product $C_1 C_2$ is also real, so that the real elements form a subalgebra $\mathcal{R}(\mathbb{C}_S)$ which is naturally isomorphic to $\mathbb{R}_S$ under the isomorphism where $1 \mapsto 1$ and $\beta_{[i]} \mapsto \beta_{[i]}$. (Anticipating this isomorphism, the same notation has been used for the generators of $\mathbb{R}_S$ and $\mathbb{C}_S$.) Many of the constructions already described for $\mathbb{R}_S$ can be generalised to $\mathbb{C}_S$ in an obvious way; for instance the body map is

$$\epsilon : \mathbb{C}_S \to \mathbb{C}$$
$$\sum_{\underline{\lambda} \in M_L} C_{\underline{\lambda}} \beta_{[\underline{\lambda}]} \mapsto C_{\emptyset}. \tag{3.27}$$

3.4   Further super matrices

Super matrices were introduced in Section 2.3. In this section the entries in a super matrix will be elements of some Grassmann algebra. Some of the results proved here are valid for more general super algebras, but may of the proofs presented here use the body map, both because this leads to simpler proofs and because the methods are well adapted to supermanifolds.

There is a simple criterion for the invertibility of a super matrix $\mathcal{M}$ with Grassmann entries; it is that the matrix is invertible if and only if the body $\epsilon(\mathcal{M})$ of $\mathcal{M}$, that is, the $(p+q) \times (p+q)$ real (or complex) matrix each of whose elements is the body of the corresponding element of $\mathcal{M}$, is invertible. This may be proved by explicitly constructing the inverse in the following manner. Let

$$\mathcal{N} = (\epsilon(\mathcal{M}))^{-1} (I - s(\mathcal{M})\epsilon(\mathcal{M})^{-1} + \left(s(\mathcal{M})\epsilon(\mathcal{M})^{-1}\right)^2 -) \dots \tag{3.28}$$

where $s(\mathcal{M})$ is the $(p, q) \times (p, q)$ super matrix whose elements are the souls of the elements of $\mathcal{M}$. Then explicit calculation shows that $\mathcal{M}\mathcal{N} = \mathcal{N}\mathcal{M} = I$ as required. This shows that a super matrix $\mathcal{M}$ is invertible if and only if both $\mathcal{M}_{0,0}$ and $\mathcal{M}_{1,1}$ are invertible.

This criterion for invertibility is used in the proof of the following lemma, which shows that the dimension of a super $\mathbb{A}$-module is well defined.

Lemma 3.4.1 *Suppose that* $\mathbb{V}$ *is a free super* $\mathbb{A}$ *module and that* $\{C_1, \ldots, C_{m+n}\}$ *is a* (m, n) *super basis and* $\{B_1, \ldots, B_{p+q}\}$ *a* (p, q) *super basis of* $\mathbb{V}$. *Then* $m = p$ *and* $n = q$.

Proof Let $\mathcal{M}$ be the $(m, n) \times (p, q)$-dimensional super matrix with entries in $\mathbb{A}$ which expresses the elements of the first super basis in terms of the second. Also let $\mathcal{N}$ be the $(p, q) \times (m, n)$ super matrix which expresses the elements of the second super basis in terms of the first. Then

$$\mathcal{N}\mathcal{M} = I_{p,q} \quad \text{and} \quad \mathcal{M}\mathcal{N} = I_{m,n}. \tag{3.29}$$

Hence

$$\epsilon(\mathcal{M})\epsilon(\mathcal{N}) = 1_{m+n} \quad \text{and} \quad \epsilon(\mathcal{N})\epsilon(\mathcal{M}) = 1_{p+q}, \tag{3.30}$$

and thus by standard arguments $m + n = p + q$. Additionally, $\epsilon(\mathcal{M})$ and $\epsilon(\mathcal{N})$ are invertible square matrices with real (or complex) entries. Now suppose that $m < p$ (and thus that $n > q$). Then $\epsilon(\mathcal{N})$ has the block form

$$\begin{pmatrix} T_1 & 0 & 0 \\ T_2 & 0 & 0 \\ 0 & T_3 & T_4 \end{pmatrix}$$

where T_1 is an $m \times m$ matrix, T_2 is a $(p - m) \times m$ matrix, T_3 is a $q \times (n - q)$ matrix and T_4 is a $q \times q$ matrix. Thus $\det(\epsilon(\mathcal{N})) = 0$, but this contradicts the invertibility of $\epsilon(\mathcal{N})$. Thus $m \geq p$. Similarly it can be shown that $p \geq m$, and thus $m = p$ as required.   ∎

The notion of determinant has many applications in differential geometry; for supermanifolds the corresponding notion is the superdeterminant, first defined by Berezin [16], which is not simply a straightforward generalisation of the conventional determinant. One may, by considering simple examples, readily see that no simple generalisation of the usual definition preserves the product property of the determinant; also, if one considers the related concept of trace one finds that the cyclic property $\text{tr}AB = \text{tr}BA$ is not valid for super matrices. However, by modifying the definition of trace the cyclic property is regained and a corresponding modification of the determinant suggested.

Definition 3.4.2 Suppose that $\mathcal{M} = \begin{pmatrix} M_{0,0} & M_{0,1} \\ M_{1,0} & M_{1,1} \end{pmatrix}$ is a square super matrix. Then the *supertrace* of $\mathcal{M}$, denoted str$\mathcal{M}$, is defined by

$$\text{str}\mathcal{M} = \text{tr}\mathcal{M}_{0,0} - \text{tr}\mathcal{M}_{1,1}. \tag{3.31}$$

The key property of the supertrace, established in the following theorem, is that it has the cyclic property. This means that an endomorphism of free super algebra module has a supertrace which is independent of the choice of super basis. (A more canonical definition of supertrace in terms of eigenvalues is also possible.)

Theorem 3.4.3 *Let $\mathcal{M}$ and $\mathcal{N}$ be $(p,q) \times (p,q)$ super matrices. Then*

$$\text{str}\mathcal{MN} = \text{str}\mathcal{NM}. \tag{3.32}$$

Proof

$$\text{str}MN = \sum_{i=1}^{p}\sum_{j=1}^{p+q}\mathcal{M}_{ij}\mathcal{N}_{ji} - \sum_{i=1}^{q}\sum_{j=1}^{p+q}\mathcal{M}_{i+p\,j}\mathcal{N}_{j\,i+p}$$

$$= \sum_{i=1}^{p}\sum_{j=1}^{p+q}\mathcal{N}_{ij}\mathcal{M}_{ji} - \sum_{i=1}^{q}\sum_{j=1}^{p+q}\mathcal{N}_{i+p\,j}\mathcal{M}_{j\,i+p}$$

$$= \text{str}NM. \tag{3.33}$$

∎

It is possible to define the superdeterminant of a square super matrix as the exponential of the supertrace of the logarithm of the super matrix. However, while this approach easily leads to the product property, it involves concepts of function which will not be introduced until the next chapter. The following definition is more direct, and is clearly equivalent when the super matrix concerned is diagonal.

Definition 3.4.4 Suppose that $\mathcal{M} = \begin{pmatrix} M_{0,0} & M_{0,1} \\ M_{1,0} & M_{1,1} \end{pmatrix}$. Then the *superdeterminant* of $\mathcal{M}$, denoted sdet$\mathcal{M}$, is defined by

$$\text{sdet}\mathcal{M} = \det(\mathcal{M}_{0,0} - \mathcal{M}_{0,1}\mathcal{M}_{1,1}^{-1}\mathcal{M}_{1,0})\det\mathcal{M}_{1,1}^{-1} \tag{3.34}$$

where the determinant of a matrix whose entries are even elements of a super algebra is calculated according to the usual formula.

That this definition has the required features is far from immediately obvious. The following theorem establishes the key product property of the superdeterminant; there are many ways to prove this result, a very general proof is given by Leĭtes in [99]. The proof given here assumes that the entries in the matrices concerned are elements of a Grassmann algebra.

Theorem 3.4.5 *Suppose that $\mathcal{M}$ and $\mathcal{N}$ are invertible super matrices with entries in $\mathbb{R}_S$ or $\mathbb{C}_S$. Then*

$$\mathrm{sdet}\mathcal{M}\mathcal{N} = \mathrm{sdet}\mathcal{M}\,\mathrm{sdet}\mathcal{N}. \qquad (3.35)$$

Proof The result will be proved by induction over the number of Grassmann generators. It must be shown that

$$\det(\mathcal{M}_{0,0} - \mathcal{M}_{0,1}\mathcal{M}_{1,1}^{-1}\mathcal{M}_{1,0})\det\mathcal{M}_{1,1}^{-1}\det(\mathcal{N}_{0,0} - \mathcal{N}_{0,1}\mathcal{N}_{1,1}^{-1}\mathcal{N}_{1,0})\det\mathcal{N}_{1,1}^{-1}$$
$$= \det((\mathcal{M}\mathcal{N})_{0,0} - (\mathcal{M}\mathcal{N})_{0,1}(\mathcal{M}\mathcal{N})_{1,1}^{-1}(\mathcal{M}\mathcal{N})_{1,0})\det(\mathcal{M}\mathcal{N})_{1,1}^{-1}. \qquad (3.36)$$

If all elements of $\mathcal{M}$ and $\mathcal{N}$ are in the subalgebra $\mathbb{R}_{S[0]} \cong \mathbb{R}$ of $\mathbb{R}_S$ then (3.36) reduces to a simple relation between ordinary determinants which is clearly true. Assume as inductive hypothesis that the result is true whenever $\mathcal{M}$ and $\mathcal{N}$ have entries in the subalgebra $\mathbb{R}_{S[L-1]}$ of $\mathbb{R}_S$. Now suppose that $\mathcal{M}$ and $\mathcal{N}$ have entries in $\mathbb{R}_{S[L]}$, so that

$$\mathcal{M} = \mathcal{P}_{L-1}\mathcal{M} + \beta_{[L]}\mathcal{A}$$
$$\mathcal{N} = \mathcal{P}_{L-1}\mathcal{N} + \beta_{[L]}\mathcal{B} \qquad (3.37)$$

where $\mathcal{P}_{L-1}$ denotes the projection of $\mathbb{R}_{S[L]}$ onto $\mathbb{R}_{S[L-1]}$ which maps $\beta_{[L]}$ to zero and each other generator to itself, and $\mathcal{A}$ and $\mathcal{B}$ are $(p+q) \times (p+q)$ matrices with entries in $\mathbb{R}_{S[L-1]}$.

Using the result that, if A is an invertible square matrix with even entries and B, also square and of the same order, has odd entries,

$$\det(A + \beta_{[L]}B) = \det A(1 + \beta_{[L]}B_i{}^j(A^{-1})_j{}^i), \qquad (3.38)$$

explicit calculation shows that equation (3.36) holds when $\mathcal{M}$ and $\mathcal{N}$ have entries in $\mathbb{R}_{S[L]}$. Hence the theorem is proved by induction. (The proof in the complex case is essentially the same.) ∎

In Chapter 9 it will be shown that $\text{GL}(p, q; \mathbb{R}_S)$ provides an example of a super Lie group, and that various of its subgroups (such as the subgroup consisting of matrices whose superdeterminants all equal 1) are also super Lie groups.

The presence of anticommuting elements requires a modification to the definition of transpose (but not to Hermitian conjugate), because of the multiplicative property $(ab)^* = b^* a^*$ specified in (3.21) for elements of a complex Grassmann algebra). If $\mathcal{M}$ is a $(p, q) \times (r, s)$ super matrix with block form $\begin{pmatrix} A & B \\ C & D \end{pmatrix}$, then its super transpose is the $(r, s) \times (p, q)$ super matrix

$$\mathcal{M}^{ST} = \begin{pmatrix} A^T & -C^T \\ B^T & D^T \end{pmatrix} \tag{3.39}$$

where A^T is the standard transpose of A and so on. This super transpose has the important property that

$$(\mathcal{M}\mathcal{N})^{ST} = \mathcal{N}^{ST} \mathcal{M}^{ST}, \tag{3.40}$$

which leads to super analogues of the orthogonal groups, known as orthosymplectic groups, as described in Section 9.2.

If the super matrix $\mathcal{M}$ has entries in $\mathbb{C}_S$, then its super hermitian conjugate is simply the matrix $\mathcal{M}^\dagger$ with entries $\mathcal{M}^\dagger_{ij} = (\mathcal{M}_{ji})^*$, that is, the standard Hermitian conjugate. Here again the product rule is as expected, that is,

$$(\mathcal{M}\mathcal{N})^\dagger = \mathcal{N}^\dagger \mathcal{M}^\dagger \tag{3.41}$$

which leads naturally to the existence of super unitary groups.

Chapter 4

Functions of anticommuting variables

This chapter develops the analysis of functions on the superspace $\mathbb{R}_S^{m,n}$ (and on some other superspaces described in Chapter 2). Since supermanifolds are spaces locally modelled on some superspace, it is crucial to decide the appropriate concept of 'superdifferentiation' and 'supersmoothness' for functions on such spaces, so that a satisfactory criterion can be applied to transition functions between overlapping coordinate patches. (If one takes the alternative, algebro-geometric, approach to manifolds described in Chapter 8, this question takes a different but closely related form.) In the heuristic approach to superspace developed in the physics literature by Volkov and Akulov [152] and Salam and Strathdee [135], a function on (m, n)-dimensional superspace is referred to as a superfield (as is explained more fully in Section 13.1) and takes the form

$$f(x^1, \ldots, x^m; \xi^1, \ldots, \xi^n) = \sum_{\underline{\mu} \in M_n} f_{\underline{\mu}}(x)\xi^{\underline{\mu}} \qquad (4.1)$$

where the $\underline{\mu}$ are multi indices in the set M_n (defined in Section 3.1) and, if $\underline{\mu} = \mu_1 \ldots \mu_k$, then $\xi^{\underline{\mu}} = \xi^{\mu_1} \ldots \xi^{\mu_k}$, while the $f_{\underline{\mu}}$ are 'ordinary functions' - a notion that in fact needs elucidation, and may be interpreted in a number of ways. Now all definitions of functions on superspace have exactly this dependence on the odd variables $\xi_1, \ldots, \xi_n$, but the nature of the functions $f_{\underline{\mu}}$ of the even variables depends on (among other things) the choice of Grassmann algebra which is made.

In Section 4.3 supersmooth functions on the superspace $\mathbb{R}_S^{m,n}$ built from the infinite-dimensional Grassmann algebra $\mathbb{R}_S$ are defined. These functions are referred to as G^∞ functions, and are much used in this book. The definition used is algebraic, and makes use of the concept of Grassmann analytic continuation described in Definition 4.2.2, a procedure for extending a function of whose domain is a subset U of $\mathbb{R}^m$ or $\mathbb{C}^m$ to a function whose domain is the set $(\epsilon_{m,0})^{-1}(U)$ of even Grassmann variables. The properties of these functions are described in Section 4.4. Grassmann analytic continuation also suggests a more restricted class of functions which can be defined on the various superspaces considered. These functions, referred to as H^∞ functions, are crucial to understanding the link between algebro-geometric and concrete supermanifolds. In terms of the superfield expansion (4.1), the distinction between H^∞ and G^∞ functions is, informally, that in the H^∞ case the 'ordinary' functions $f_{\underline{\mu}}$ take values in $\mathbb{R}$, while in the G^∞ case they take values in $\mathbb{R}_S$.

Sections 4.3 and 4.4 are essential to the development of supermanifolds in this book, while the other parts of the chapter are included to give a more complete account of different approaches which may be taken, and because certain theorems for infinite dimensional super algebras are most readily proved by using truncation to finite-dimensional algebras.

Partly to motivate the definition of G^∞ function on $\mathbb{R}_S^{m,n}$, but also for application where truncations to a finite-dimensional Grassmann algebra is required, a description is given in Section 4.1 of the notion of super-smooth functions on the superspace $\mathbb{R}_{S[L]}^{m,n}$ which is built from the finite-dimensional Grassmann algebra $\mathbb{R}_{S[L]}$, and the theory further developed in Section 4.2 on Taylor's theorem and Grassmann analytic continuation. In the finite dimensional setting a notion of differentiation in terms of small increments can be developed. However odd derivatives are ambiguous, a problem which can be dealt with in two different ways; one is to use infinite-dimensional algebras, which leads to the key definition of G^∞ functions on $\mathbb{R}_S^{m,n}$. This definition uses the idea of Grassmann analytic continuation for finite-dimensional Grassmann algebras as the definition of a smooth function when an infinite-dimensional algebra is involved. The other approach, described in Section 4.6, is to slightly restrict the class of supersmooth functions, considering a class of functions referred to as GH^∞ functions.

Some other infinite-dimensional Grassmann algebras are discussed in Section 4.5, and then two sections consider partitions of unity and the inverse function theorem. The final section considers superholomorphic functions on $\mathbb{C}_S^{m,n}$.

Further aspects of analysis on superspace have been considered by various authors, including for example a study of distributions and partial differential equations by Khrennikov [87].

4.1 Superdifferentiation and finite-dimensional Grassmann algebras

Using the finite-dimensional Grassmann algebra $\mathbb{R}_{S[L]}$ as our super commutative algebra (with $\mathbb{R}_{S[L]}^{m,n}$ as the corresponding superspace) there are two approaches one can take in defining a notion of differentiability; one is to seek some direct analogue of the usual notion by considering changes in the function corresponding to small changes in its argument [118], while the other is to extend smooth functions from $\mathbb{R}^m$ to $\mathbb{R}_{S[L]}^{p,0}$ by Taylor expansion [43, 118]; both these approaches will now be described, and the relation between them explained. To begin with the differentiability property is defined, leading to supersmooth functions which are referred to as G^∞ functions, to emphasize the analogy with conventional C^∞ functions.

Definition 4.1.1 Let U be an open set in $\mathbb{R}_{S[L]}^{m,n}$ and $f : U \to \mathbb{R}_{S[L]}$. Then

(a) f is said to be G^0 if f is continuous on U with respect to the usual finite-dimensional vector space topology.

(b) f is said to be G^1 on U if there exist m continuous functions $\partial_i^E f : U \to \mathbb{R}_{S[L]}, i = 1, \ldots, m$, and n continuous functions $\partial_j^O f : U \to \mathbb{R}_{S[L]}, j = 1, \ldots, n$ and a function $\rho : \mathbb{R}_{S[L]}^{m,n} \to \mathbb{R}_{S[L]}$ which satisfies

$$\|\rho(h; \eta)\| \to 0 \quad \text{as} \quad \|(h; \eta)\| \to 0 \tag{4.2}$$

such that, if $(x; \xi)$ and $(x + h; \xi + \eta)$ are both in U, then

$$f(x + h; \xi + \eta) = f(x; \xi) + \sum_{i=1}^{m} h^i (\partial_i^E f)(x; \xi)$$

$$+ \sum_{j=1}^{n} \eta^j (\partial_j^O f)(x; \xi) + \|(x; \xi)\| \rho(h; \eta). \tag{4.3}$$

(Since $\mathbb{R}_{S[L]}$ is not a field, the odd partial derivatives $\partial_j^O f, j = 1, \ldots, n$ will not in general be unique.)

(c) The definition of G^p, where p is a finite positive integer, is made inductively. A function f is said to be G^p on U if f is G^1 on U and it is possible to choose $\partial_k^S f, k = 1, \ldots, m + n$ which are G^{p-1} on U. (Here ∂_k^S denotes a partial derivative of either parity, with $\partial_i^S = \partial_i^E$ for $i = 1, \ldots, m$ and $\partial_{j+m}^S = \partial_j^O$ for $j = 1, \ldots, n$).

(d) f is said to be G^∞ on U if f is G^p on U for any positive integer p.

(e) f is said to be G^ω on U if, given any point $X = (x; \xi)$ in U, there exists a neighbourhood N_X of X such that, for all $Y = (y, v)$ in N_X, $f(Y)$ is equal to the sum of an absolutely convergent power series of this form:

$$f(X) = \sum_{k_1=0,\ldots,k_{m+n}=0}^{\infty} a_{k_1\ldots k_{m+n}}(Y^1 - X^1)^{k_1} \ldots (Y^{m+n} - X^{m+n})^{k_{m+n}}$$

(4.4)

(with each coefficient $a_{k_1\ldots k_{m+n}}$ in $\mathbb{R}_{S[L]}$).
(f) The set of G^p functions of U into $\mathbb{R}_{S[L]}$ is denoted $G^p(U)$.

Also suppose that $g : U \to \mathbb{R}^{r,s}_{S[L]}$. Then g is said to be G^∞ on U if each of the $r + s$ components of g is G^∞. The set of all such functions is denoted $G^\infty(U, \mathbb{R}^{r,s}_{S[L]})$.

Example 4.1.2   A simple example of a G^∞ function is

$$f : \mathbb{R}^{2,2}_S \to \mathbb{R}_{S[L]}$$
$$(x^1, x^2; \xi^1, \xi^2) \mapsto cx^1(x^2)^2\xi^1\xi^2$$

(4.5)

(where c is some fixed element of $\mathbb{R}_{S[L]0}$). By 'differentiation from first principles' one calculates that

$$\partial^E_1 f(x^1, x^2; \xi^1, \xi^2) = c(x^2)^2\xi^1\xi^2$$
$$\partial^E_2 f(x^1, x^2; \xi^1, \xi^2) = 2cx^1 x^2\xi^1\xi^2$$
$$\partial^O_1 f(x^1, x^2; \xi^1, \xi^2) = cx^1(x^2)^2\xi^2$$
$$\partial^O_2 f(x^1, x^2; \xi^1, \xi^2) = -cx^1(x^2)^2\xi^1 .$$

(4.6)

The definition of superdifferentiability bears some resemblance to the concept of analyticity of functions of complex variables, but there are important differences; for instance, while it can readily be established that a G^ω function is always G^∞, the converse is not true (as will emerge in the following section).

It can be shown that a G^∞ function is always C^∞ [118], but again the converse is not true (this time, as one might expect from the analogy of complex analytic function theory). In fact a G^∞ function is a C^∞ function whose differential is not merely linear with respect to $\mathbb{R}$, but also linear with respect to the algebra $\mathbb{R}_{S[L]}$. (This has been used as a definition of superdifferentiability by Jadzcyk and Pilch [84].) A theory of differential calculus for finite dimensional Grassmann algebras was also introduced by Vladimir and Volovich[150].

4.2 Taylor expansion and Grassmann analytic continuation

This section begins with a theorem for Taylor expansion of G^∞ functions. While this theorem bears a close resemblance to the conventional Taylor's theorem for C^∞ functions, situations where the increments are nilpotent provide finite Taylor series which are exact, and require no remainder term. This makes possible an algebraic characterisation of G^∞ functions, and also leads to some alternative classes of supersmooth functions.

Theorem 4.2.1 *Suppose that U is an open set in $\mathbb{R}^{m,n}_{S[L]}$ and f is a function in $G^\infty(U)$. Also suppose that $(x;\xi)$ and $(h;\eta)$ are elements of $\mathbb{R}^{m,n}_{S[L]}$ such that, for $0 \le t \le 1$, the point $(x + th; \xi + t\eta)$ lies in U. Then*

$$f(x + h; \xi + \eta) =$$

$$\sum_{k_1=0...k_{m+n}=0}^{k_1=r,k_2=r-k_1,...,k_{m+n}=r-(k_1+...k_{m+n-1})} \left[\frac{1}{k_1!...k_{m+n}!}(H^1)^{k_1}...(H^{m+n})^{k_{m+n}} \right.$$

$$\times \quad (\partial^S_{m+n})^{k_{m+n}}...(\partial^S_1)^{k_1}f(x;\xi) \Big]$$

$$+ \sum_{\substack{k_1=0...k_{m+n}=0 \\ k_1+\cdots+k_{m+n}=p+1}} \left[\frac{1}{k_1!...k_{m+n}!}(H^1)^{k_1}..(H^{m+n})^{k_{m+n}} \right.$$

$$\times \quad \int_0^t (\partial^S_{m+n})^{k_{m+n}}...(\partial^S_1)^{k_1}f(x + th; \xi + t\eta)dt \Big], \quad (4.7)$$

where $H^i = h^i, i = 1,...,m$ and $H^{j+m} = \eta^j, j = 1,...,n$.

Proof As in the classical case this theorem may be proved by setting

$$u : [0,1] \to \mathbb{R}_{S[L]},$$

$$t \mapsto f(x + th; \xi + t\eta) \quad (4.8)$$

and integrating the identity

$$\frac{d}{dt}(u(t) + (1-t)u'(t) + \cdots + \frac{1}{p!}(1-t)^p u^{(p)}(t)) = \frac{1}{p!}(1-t)^p u^{(p+1)}(t) \quad (4.9)$$

between limits 0 and 1. ∎

This theorem will be used to give a classification of G^∞ functions in terms of C^∞ functions. In order to do this, Taylor expansion is used to define a natural extension, referred to as *Grassmann analytic continuation* of functions on $\mathbb{R}^m$ to $\mathbb{R}_S^{m,0}$ and denoted by the symbol $\hat{}$ over the function. (In general any restriction in some sense to the 'body' will be denoted by a subscript $\emptyset$ while a lift (provided there is no ambiguity) will be denoted by the symbol $\hat{}$ over the object concerned.)

Definition 4.2.2   Suppose that V is an open subset of $\mathbb{R}^m$ and U is a subset of $\mathbb{R}_{S[L]}^{m,n}$ such that $\epsilon_{m,n}(U) = V$. (Recall from equation (3.6) that $\epsilon_{m,n}$ is the projection map taking a point in $\mathbb{R}_{S[L]}^{m,n}$ onto its body in $\mathbb{R}^m$.) Then

$$\hat{} : C^\infty(V, \mathbb{R}_{S[L]}) \to \big\{ \text{functions of } U \text{ into } \mathbb{R}_{S[L]} \big\} \qquad (4.10)$$

is defined by

$$\widehat{f}(x;\xi) =_{def}$$

$$\sum_{i_1=0,\ldots,i_m=0}^{L} \frac{1}{i_1! \ldots i_m!} \partial_1^{i_1} \ldots \partial_m^{i_m} f(\epsilon_{m,n}(x)) \times s(x^1)^{i_1} \ldots s(x^m)^{i_m} .$$

$$(4.11)$$

Given f in $C^\infty(V, \mathbb{R}_{S[L]})$, the function $\widehat{f}$ is called the *Grassmann analytic continuation* of f.

This definition can be applied to a C^∞ function whose codomain is a subset of $\mathbb{R}_{S[L]}$, and to a set with a natural inclusion in $\mathbb{R}_{S[L]}$; a case which occurs frequently is the continuation of a real-valued C^∞ function. function on $\mathbb{R}^m$ to $\mathbb{R}_{S[L]}^{m,n}$ is G^∞ and that their derivatives correspond in a natural way.

Theorem 4.2.3   *Suppose that U is an open subset of $\mathbb{R}_{S[L]}^{m,n}$, and that f and g are functions in $C^\infty(\epsilon_{m,n}(U), \mathbb{R}_{S[L]})$, the space of C^∞ maps of $\epsilon_{m,n}(U)$ into $\mathbb{R}_{S[L]}$ regarded simply as a Banach space over the reals. Then*

(a) $\widehat{f} \in G^\infty(U)$,
(b) $\partial_i^E(\widehat{f}) = \widehat{\partial_i f}$ for $i = 1, \ldots, m$,
(c) $\partial_{j+n}^O(\widehat{f}) = 0$ for $j = 1, \ldots, n$, and
(d) $\widehat{f+g} = \widehat{f} + \widehat{g}$ and $\widehat{fg} = \widehat{f}\,\widehat{g}$.

Proof (c) is immediate since the dependence on odd variables is trivial, and the proof of (a) and (b) can be restricted to the case where $n = 0$.

Let (x) and $(x + h)$ be in U. Then

$$\widehat{f}(x + h) = \widehat{f}(\epsilon_{m,0}(x + h))$$

$$+ \sum_{i_1=0,\dots,i_m=0}^{L} \frac{1}{i_1!\dots i_m!} \partial_1^{i_1} \dots \partial_m^{i_m} f(\epsilon_{m,0}(x + h))$$

$$\times s(x^1 + h^1)^{i_1} \dots s(x^m + h^m)^{i_m}. \tag{4.12}$$

Also the classical Taylor's theorem gives that

$$f(\epsilon_{m,0}(x + h)) = f(\epsilon_{m,0}(x))$$

$$+ \sum_{i_1=0,\dots,i_m=0}^{i_1=L,i_2=L-i_1,\dots,i_m=L-(i_1+\dots+i_{m-1})} \left[\frac{1}{i_1!\dots i_m!} \partial_1^{i_1} \dots \partial_m^{i_m} f(\epsilon_{m,0}(x)) \right.$$

$$\times \epsilon_{m,0}(h^1)^{i_1} \dots \epsilon_{m,0}(h^m)^{i_m} \Big]$$

$$\sum_{\substack{i_1=0\dots i_m=0 \\ i_1+\dots+i_m+n=L+1}} \left[\frac{1}{i_1!\dots i_m!} \epsilon_{m,0}(h^1)^{i_1}_{\cdot\cdot} \epsilon_{m,0}(h^m)^{i_m} \right.$$

$$\times \int_0^t (\partial_{m+n}^S)^{i_m} \dots (\partial_1^S)^{i_1} f(x + th) dt \Big]. \tag{4.13}$$

Now the algebraic property of the classical Taylor expansion obtained from equating the expressions for $f(a + b + c)$ obtained in two steps and in one (with $a = \epsilon_{m,0}(x)$ $b = \epsilon_{m,0}(h)$ and $c = s(x + h)$) shows that $\widehat{f}$ satisfies the Taylor theorem 4.2.1 and thus has first derivatives satisfying (b). It also follows by induction that $\widehat{f}$ is G^∞, so that (a) is established. Finally (d) follows from the fact that the classical Taylor expansion of the sum of two functions is the sum of the Taylor expansions of the functions, and the Taylor expansion of a product is the product of the Taylor expansions. ∎

The following theorem shows that the coefficient functions f_μ in the super-field expansion (4.1) may be understood as Grassmann analytic continuations of C^∞ functions on $\mathbb{R}^m$. The Grassmann analytic continuation also allows one to define other useful classes of supersmooth functions.

Theorem 4.2.4   *Given a function f in $G^\infty(U)$, there exist functions $f_{\underline{\mu}}$ in $C^\infty(\epsilon_{m,n}(U)), \underline{\mu} \in M_n$ such that*

$$f = \sum_{\underline{\mu} \in M_n} \widehat{f_{\underline{\mu}}} \xi^{\underline{\mu}} \tag{4.14}$$

where $\xi^j, j = 1, \ldots, n$ are odd coordinate functions

$$\xi^j(x; \xi) = \xi^j \tag{4.15}$$

and $\xi^{\underline{\mu}} = \xi^{\mu_1} \ldots \xi^{\mu_k}$ if $\underline{\mu} = \mu_1, \ldots \mu_k$.

Proof   Applying Taylor's theorem 4.2.1 with $(x; \xi)$ replaced by $(\epsilon(x^1)1, \ldots, \epsilon(x^m)1; 0 \ldots, 0)$ and $(h; \eta)$ by $(s(x^1), \ldots, s(x^m); \theta^1, \ldots, \theta^n)$ leads to (4.14), with the coefficient functions $f_{\underline{\mu}} : \epsilon_{m,n}(U) \to \mathbb{R}_{S[L]}$ defined by

$$f_{\underline{\mu}}(t^1, \ldots, t^m) = \partial^O_{\mu_k} \ldots \partial^O_{\mu_1} f(t^1 1, \ldots, t^m 1). \tag{4.16}$$

Corollary 4.2.5   *The set $G^\infty(U)$ exactly coincides with the set of functions $f : U \to \mathbb{R}_{S[L]}$ which can be expressed in the form (4.14) for some functions $f_{\underline{\mu}}$ in $C^\infty(\epsilon U, \mathbb{R}_{S[L]})$.*

This corollary provides an alternative characterisation of G^∞ functions which could equally well have been used as a definition, without recourse to any topology other than the DeWitt topology on the Grassmann algebra. This approach will be used in the following sections to define a notion of supersmoothness when the Grassmann algebra has an infinite number of generators, and to define some alternative, more restricted, classes of supersmooth functions.

4.3   Supersmooth functions on $\mathbb{R}^{m,n}_S$

In this section supersmooth or G^∞ functions on the infinite-dimensional space $\mathbb{R}^{m,n}_S$ will be defined. This definition underpins the concrete supermanifold constructions considered in this book. (Using an infinite-dimensional Grassmann algebra resolves the difficulties of ambiguous odd derivatives encountered in Definition 4.1.1, where finite-dimensional Grassmann algebras are used, because there are no longer any elements which are annihilated by an arbitrary odd element of the algebra.) Many of the difficulties usually encountered when working with an infinite-dimensional vector space are avoided by the use of the DeWitt topology together with the Grassmann analytic continuation which is now defined (extending the construction of Theorem 4.2.4).

Definition 4.3.1   Let be V is open in $\mathbb{R}^m$ and $f : V \to \mathbb{R}_S$.

(a) f is said to be C^∞ if for each positive integer L the function $\mathcal{P}_L \circ f : V \to \mathbb{R}_{S[L]}$ is C^∞. (Here $\mathcal{P}_L$ denotes the projection of $\mathbb{R}_S$ onto $\mathbb{R}_{S[L]}$ obtained by setting all generators $\beta_{[r]}$ with $r > L$ to zero.) The set of all such functions is denoted $C^\infty(V, \mathbb{R}_S)$.

(b) If f is a function in $C^\infty(V, \mathbb{R}_S)$ then the function $\widehat{f} : (\epsilon_{m,0})^{-1}(V) \to \mathbb{R}_S$ is defined by

$$\widehat{f}(x; \xi)$$

$$=_{def} \sum_{i_1=0,\ldots,i_m=0}^{\infty} \frac{1}{i_1! \ldots i_m!} \partial_1^{i_1} \ldots \partial_m^{i_m} f(\epsilon_{m,n}(x)) \times s(x^1)^{i_1} \ldots s(x^m)^{i_m}.$$

$$(4.17)$$

This Grassmann analytic continuation has the properties (b) and (c) of Theorem 4.2.3 for the finite-dimensional case, as may be seen by considering terms involving the first $1, 2, \ldots$ Grassmann generators in turn.

The basic idea of a supersmooth function is that it should have a superfield expansion of the form (4.1) where the coefficient functions f_μ are Grassmann analytic continuations of C^∞ functions.

Definition 4.3.2   Let U be open in $\mathbb{R}_S^{m,n}$. Then $f : U \to \mathbb{R}_S$ is said to be G^∞ on U if and only if there exists a collection $\left\{ f_\mu | \mu \in M_n \right\}$ of functions in $C^\infty(\epsilon_{m,n}(U))$ such that

$$f(x; \xi) = \sum_{\mu \in M_n} \widehat{f_\mu}(x) \xi^\mu \qquad (4.18)$$

for each $(x; \xi)$ in U. This expansion is called the Grassmann analytic expansion of f and the functions f_μ are called the Grassmann analytic coefficients of f.

As the notation suggests, G^∞ functions are infinitely differentiable, with differentiation defined in the following manner:

Definition 4.3.3   Suppose that f is in $G^\infty(U)$ with Grassmann expansion coefficients f_μ. Then for $i = 1, \ldots, m$ the derivative $\partial_i^E f$ is defined to be the function in $G^\infty(U)$ with Grassmann expansion coefficients $\partial_i f_\mu$

while for $j = 1, \ldots, n$ the derivative $\partial_j^O f$ is defined by

$$\partial_j^O f(x; \xi) = \sum_{\underline{\mu} \in M_n} (-1)^{|f_{\underline{\mu}}|} p_{j,\underline{\mu}} \, \widehat{f_{\underline{\mu}}}((x)\xi^{\underline{\mu}/j}) \qquad (4.19)$$

$$\text{where} \quad p_{j,\underline{\mu}} = \begin{cases} (-1)^{\ell+1} & \text{if } j = \mu_\ell \\ 0 & \text{otherwise} \end{cases}$$

$$\text{and} \quad \underline{\mu}/j = \begin{cases} \mu_1 \ldots \mu_{\ell-1} \mu_{\ell+1} \ldots \mu_k & \text{if } j = \mu_\ell \\ 0 & \text{otherwise} \end{cases}$$

Although it is necessary to use an infinite number of generators to obtain well-defined odd derivatives, it is often useful to truncate to a finite number of generators when proving theorems about G^∞ functions. If H is any function of an open subset U of $\mathbb{R}_S^{m,n}$ into $\mathbb{R}_S$, then the notation $H_{[L]}$ will be used to denote the truncated function $p_{[L]} \circ H : U_{[L]} \to \mathbb{R}_{S[L]}$, where $U_{[L]} = p_{[L]}^{m,n}(U)$, $\mathcal{P}_L$ is again the projection of $\mathbb{R}_S$ onto $\mathbb{R}_{S[L]}$ (and $\mathcal{P}_{[L]}^{m,n}$ the corresponding projection of $\mathbb{R}_S^{m,n}$ onto $\mathbb{R}_{S[L]}^{m,n}$). It is then an immediate consequence of Theorems 4.2.3 and 4.2.4 and Corollary 4.2.5 that a function H on an open subset of $\mathbb{R}_S^{m,n}$ is G^∞ (according to Definition 4.3.1) if and only if the truncated function $H_{[L]}$ is G^∞ (according to Definition 4.1.1) for every positive integer L.

4.4 Properties of supersmooth functions

In this section it is shown that many standard results in the analysis of conventional, smooth functions have their analogues for supersmooth functions on the superspace $\mathbb{R}_S^{m,n}$. A graded Leibniz rule (for the differentiation of a product) and a chain rule for differentiating a function of a function will be established, and it will be shown that the various function spaces (such as $G^\infty(U)$) are super algebras, and the structure of these algebras will be discussed. The first theorem establishes a number of key properties of G^∞ functions.

Theorem 4.4.1 *Let a, b be elements of $\mathbb{R}_S$, U be open in $\mathbb{R}_S^{m,n}$, and f and g be functions in $G^\infty(U)$. Then*

(a) The function $f + g$ is in $G^\infty(U)$, and

$$\partial_k^S (f + g) = \partial_k^S f + \partial_k^S g \qquad i = 1, \ldots, m + n. \qquad (4.20)$$

(b) bf is in $G^\infty(U)$ and

$$\partial_k^S(bf) = (-1)^{|b||k|}b\partial_k^S f, \qquad i = 1,\ldots,m+n. \qquad (4.21)$$

(c) If E and Q represent projections of $\mathbb{R}_S$ onto its even and odd parts respectively, then $f_0 = E \circ f$ and $f_1 = Q \circ f$ are both in $G^\infty(U)$.
(d) The function fg is in $G^\infty(U)$, with

$$\partial_k^S(fg) = (\partial_k^S f)g + (-1)^{|f||k|}f(\partial_k^S g), k = 1,\ldots m+n. \qquad (4.22)$$

(e) $G^\infty(U)$ is a super algebra with the product defined pointwise and

$$G^\infty(U)_0 = \{f|f \in G^\infty(U), f(U) \subset \mathbb{R}_{S0}\},$$
$$G^\infty(U)_1 = \{f|f \in G^\infty(U), f(U) \subset \mathbb{R}_{S1}\}. \qquad (4.23)$$

Proof Suppose that, for μ in M_n, the functions $f_\mu : \epsilon_{m,n}(U) \to \mathbb{R}_S$ are the Grassmann expansion coefficients of f, and g_μ are the Grassmann expansion coefficients of g.

(a) For all $(x;\xi)$ in U

$$(f+g)(x;\xi) = \sum_{\mu\in M_n}(\widehat{f_\mu}(x) + \widehat{g_\mu}(x))\xi^\mu$$

$$= \sum_{\mu\in M_n}(\widehat{f_\mu + g_\mu})(x)\xi^\mu \qquad (4.24)$$

by Theorem 4.2.3(d). Hence $f + g$ is G^∞ with Grassmann analytic coefficients $f_\mu + g_\mu$; it then follows directly from Definition 4.3.3 that

$$\partial_k^S(f+g) = \partial_k^S f + \partial_k^S g \qquad k = 1,\ldots,m+n. \qquad (4.25)$$

(b) Since

$$(bf)(x;\xi) = \widehat{bf_\mu}(x)\xi^\mu), \qquad (4.26)$$

bf is G^∞, with Grassmann analytic coefficients bf_μ, so that

$$\partial_i^E(bf) = b\partial_i^E f \qquad i = 1,\ldots,m$$
$$\partial_\lambda^O(bf) = (-1)^{|b|}\partial_\lambda^S f \qquad \lambda = m+1,\ldots,m+n \qquad (4.27)$$

as required.

(c) Since

$$f_0(x;\xi) = \sum_{\underline{\mu}\in M_{n0}} \widehat{E \circ f_{\underline{\mu}}}(x)\xi^{\underline{\mu}} + \sum_{\underline{\mu}\in M_{n1}} \widehat{Q \circ f_{\underline{\mu}}}(x)\xi^{\underline{\mu}}. \qquad (4.28)$$

and each $E \circ f_{\underline{\mu}}$ and $Q \circ f_{\underline{\mu}}$ is C^∞ (because E and Q both commute with $\mathcal{P}_L$) f_0 and f_1 are G^∞.

(d)

$$fg(x;\xi) = \sum_{\underline{\mu}\in M_n}\sum_{\underline{\nu}\in M_n} (-1)^{|\theta^{\underline{\mu}}||g_{\underline{\nu}}|}\widehat{f_{\underline{\mu}}}(x)\widehat{g_{\underline{\mu}}}(x)\theta^{\underline{\mu}}\theta^{\underline{\nu}}$$

$$= \sum_{\underline{\mu}\in M_n}\sum_{\underline{\nu}\in M_n}\sum_{\underline{\rho}\in M_n} \epsilon(\underline{\mu},\underline{\nu}:\underline{\rho})(-1)^{|\theta^{\underline{\mu}}||g_{\underline{\nu}}|}\widehat{f_{\underline{\mu}}}(x)\widehat{g_{\underline{\mu}}}(x)\theta^{\underline{\rho}}$$

$$(4.29)$$

where

$$\epsilon(\underline{\mu},\underline{\nu};\rho) = \begin{cases} 1 & \text{if } \theta^{\underline{\mu}}\theta^{\underline{\nu}} = \theta^{\underline{\rho}} \\ -1 & \text{if } \theta^{\underline{\mu}}\theta^{\underline{\nu}} = -\theta^{\underline{\rho}} \\ 0 & \text{otherwise} \end{cases} \qquad (4.30)$$

Now, by Theorem 4.2.3(d), it can be seen that that

$$\widehat{f_{\underline{\mu}}}\widehat{g_{\underline{\nu}}} = \widehat{f_{\underline{\mu}}g_{\underline{\nu}}} \qquad (4.31)$$

and thus

$$fg(x;\xi) = \sum_{\underline{\rho}\in M_n} \widehat{h_{\underline{\rho}}}(x)\xi^{\underline{\rho}} \qquad (4.32)$$

with each $h_{\underline{\rho}} : \epsilon_{m,n}(U) \to \mathbb{R}_S$ uniquely defined and C^∞. Thus fg is G^∞. Also, for $i = 1, \ldots m$,

$$\partial_i^E fg(x;\xi)$$

$$= \sum_{\underline{\mu}\in M_n}\sum_{\underline{\nu}\in M_n}\sum_{\underline{\rho}\in M_n} \epsilon(\underline{\mu},\underline{\nu};\underline{\rho})(\widehat{\partial_i f_{\underline{\mu}}}(x)\widehat{g_{\underline{\nu}}}(x)\xi^{\underline{\rho}} + \widehat{f_{\underline{\mu}}}(x)\widehat{\partial_i g_{\underline{\nu}}}(x)\xi^{\underline{\rho}}$$

$$= \partial_i^E f(x;\xi)g(x;\xi) + f(x;\xi)\partial_i^E g(x;\xi) \qquad (4.33)$$

while for $\lambda = m + 1, \ldots, m + n$ explicit calculation shows that

$$\partial_\lambda^S fg = \partial_\lambda^S fg + (-1)^{|f|}f\partial_\lambda^S g \qquad (4.34)$$

∎

The next key property of G^∞ functions which will be established is the chain rule for differentiating the composition of a function $f : \mathbb{R}^{m,n}_S \to \mathbb{R}^{p,q}_S$ with a function $g : \mathbb{R}^{p,q}_S \to \mathbb{R}^{r,s}_S$, which is most readily proved by truncating to $\mathbb{R}_{S[L]}$ (with L arbitrary) and using the increment definition of differentiation, Definition 4.1.1.

Theorem 4.4.2 *Let U be open in $\mathbb{R}^{m,n}_S$ and V be open in $\mathbb{R}^{p,q}_S$. Also, let $f : U \to \mathbb{R}^{p,q}_S$ and $g : V \to \mathbb{R}^{r,s}_S$ be G^∞, with $f(U) \subset V$. Then $g \circ f : U \to \mathbb{R}^{r,s}_S$ is G^∞ and, for $k = 1, \ldots, r + s$, $j = 1, \ldots, m + n$,*

$$\partial^S_j(g^k \circ f)(x; \xi) = \sum_{k'=1}^{p+q} \partial^S_j f^{k'}(x; \xi) \partial^S_{k'} g^k(f(x; \xi)). \qquad (4.35)$$

(Here for $k = 1, \ldots, r + s$, g^k denotes the k component of $g : V \to \mathbb{R}^{r,s}_S$, and so on.)

Proof Let L be a positive integer and suppose that X and $X + H$ are in $U_{[L]}$. For each $k = 1, \ldots, r + s$,

$$g^k_{[L]} \circ f_{[L]}(X + H)$$

$$= g^k_{[L]}\left(f_{[L]}(X) + \sum_{i=1}^{m+n} H^i \partial^E_i f_{[L]}(X) + \|H\|\eta(H) \right)$$

$$= g^k_{[L]}(f_{[L]}(X))$$

$$+ \sum_{k'=1}^{p+q} \left(\sum_{i=1}^{m+n} H^i \partial^E_i f^{k'}_{[L]}(X) + \|H\|\eta^{k'}(H) \right) \partial^E_{k'} g^k_{[L]}(f_{[L]}(X))$$

$$+ \left[\left\| \sum_{i=1}^{m+n} H^i \partial^E_i f_{[L]}(X) + \|H\|\eta(H) \right\| \right.$$

$$\left. \times \eta'^k \left(\sum_{i=1}^{m+n} H^i \partial^E_i f^j_{[L]}(X) + \|H\|\eta^j(H) \right) \right]$$

$$= g^k_{[L]}(f(X)) + \sum_{k'=1}^{p+q} \sum_{i=1}^{m+n} \left(H^i \partial^E_i f^{k'}_{[L][L]}(X) \partial^S_{k'} g^k_{[L]}(f_{[L]}(X)) \right)$$

$$+ \|H\|\eta''^k(H) \qquad (4.36)$$

where η, η' and η'' are functions which all tend to zero as their arguments tend to zero. Hence $g_{[L]} \circ f_{[L]}$ is G^1 with

$$\partial_j^S(g^k{}_{[L]} \circ f_{[L]}) = \sum_{k'=1}^{p+q} \partial_j^S f^{k'}(x;\xi) \partial_{k'}^S g^k(f(x;\xi)). \qquad (4.37)$$

Thus, by induction (and Theorem 4.4.1), $g_{[L]} \circ f_{[L]}$ is G^∞ with derivative as required. Since L is arbitrary the theorem is proved. ∎

It is also useful to distinguish a more restricted class of functions, known as H^∞ functions, which provide the bridge between the concrete and the algebro-geometric approach.

Definition 4.4.3   Let U be an open subset of $\mathbb{R}^{m,n}_{S[L]}$. Then $f : U \to \mathbb{R}_{S[L]}$ is said to be H^∞ if there exists for each $\underline{\mu}$ in M_n a C^∞ function $f_{\underline{\mu}} : \epsilon_{m,n}(U) \to \mathbb{R}$ such that

$$f(x;\xi) = \sum_{\underline{\mu} \in M_n} \widehat{f_{\underline{\mu}}}(x)\xi^{\underline{\mu}} \qquad (4.38)$$

for each $(x;\xi)$ in U. The set of all such functions is denoted $H^\infty(U)$.

The important feature of this definition is that the coefficient functions $f_{\underline{\mu}}$ are real valued and not $\mathbb{R}_{S[L]}$-valued, so that although (by Theorem 4.2.4) any H^∞ function is also G^∞, the converse is not true. Indeed even the constant function

$$f : U \to \mathbb{R}_{S[L]}$$
$$(x;\xi) \mapsto \alpha \qquad (4.39)$$

where α is some fixed odd element of the Grassmann algebra $\mathbb{R}_{S[L]}$ is not H^∞, although it is obviously G^∞. The standard supersymmetry transformations in flat superspace are also G^∞ but not H^∞, as are their counterparts in supergravity. The nice property of $H^\infty(U)$ is that its algebraic properties may be defined without reference to any particular Grassmann algebra $\mathbb{R}_{S[L]}$, as will be shown below.

4.5   Other infinite-dimensional algebras

Since $\mathbb{B}_\infty$ is a Grassmann algebra, the Definition 4.1.1 of a G^∞ function can also be applied to $\mathbb{B}_\infty$-valued functions of $\mathbb{R}^{m,n}_S$, and Taylor's theorem 4.2.1

remains true. However the Grassmann analytic continuation cannot be applied to all G^∞ functions, because the Taylor series involved in the Grassmann analytic expansion can no longer be truncated after a finite number of terms, and so will not necessarily converge. It is possible to base a consistent theory of supermanifolds on such a Banach-Grassmann algebra, and considerable development of just such a theory was made by Jadczyk and Pilch [84]; a more involved construction, together with further consideration of the extension problem, was given by Hoyos, Quiros, Lamirez Mittelbrunn and De Urries [79]. A detailed study of the soul expansion for Banach Grassmann algebras was made by Pestov [110]. However the impossibility of extending all C^∞ functions has some drawbacks, and this approach will not be pursued further here. A further possibility was developed by Inoue and Maeda [83], who put a Fréchet structure on an infinite-dimensional algebra, thus allowing all smooth functions to be extended. This provided a mathematically rigorous approach to infinite-dimensional Grassmann algebras which can be used when the simpler DeWitt approach, which avoids convergence issues, proves inadequate. In this book this is not necessary, but it is likely that in some contexts this would not be the case.

4.6 Obtaining well defined odd derivatives with finite-dimensional Grassmann algebras

In Section 4.3 G^∞ functions were defined in an attempt to mimic for functions of Grassmann variables the usual concept of C^∞ function. While a cornerstone of conventional calculus, the Taylor theorem, could also be established for G^∞ functions, the ambiguity of odd derivatives prevents a complete development along the lines of conventional calculus. In this section a slightly restricted class of functions for finite-dimensional Grassmann algebras, known as GH^∞ functions, will be introduced; as the name suggests, GH^∞ functions occupy an intermediate position between G^∞ and H^∞ functions.

The basic idea of the definition is that a GH^∞ function has a Grassmann analytic expansion whose coefficient functions take their values in some subalgebra $\mathbb{R}_{S[L']}$ of $\mathbb{R}_{S[L]}$. (In the case of a H^∞ function, the coefficient functions take their values in the reals, while for a G^∞ function they take values in the full algebra $\mathbb{R}_{S[L]}$.) Before defining these new functions the necessary subalgebras will be defined. First, suppose that L, L' are positive integers with $L' < L$. Then the generators of the Grassmann algebras $\mathbb{R}_{S[L]}$

and $\mathbb{R}_{S[L']}$ will be denoted $1_{[L]}, \beta_{[L]\,1}, \ldots, \beta_{[L]\,L}$ and $1_{[L']}, \beta_{[L']\,1}, \ldots, \beta_{[L']\,L'}$ respectively. Associated maps, such as $\epsilon_{[L]}$ will also be given suffices, to indicate the particular Grassmann algebra concerned. There is a natural injection $\iota_{[L',L]} : \mathbb{R}_{S[L']} \to \mathbb{R}_{S[L]}$ which is the unique algebra homomorphism such that

$$\iota_{[L',L]}(\beta_{[L']\,i}) = \beta_{[L]\,i} \quad i = 1, \ldots, L'$$

$$\iota_{[L',L]}(1_{[L']}) = 1_{[L]}. \tag{4.40}$$

The Grassmann analytic continuation of Definition 4.2.2 allows the domain of a $\mathbb{R}_{S[L]}$-valued function to be extended from the body $\epsilon_{m,n[L]}(U)$ of some open subset U of $\mathbb{R}_{S[L]}^{m,n}$ to the whole of U; two slight variations of the Grassmann analytic continuation, which allow the extension of the domain of real-valued and $\mathbb{R}_{S[L']}$-valued functions, will now be given.

Definition 4.6.1   Let U be open in $\mathbb{R}_{S[L]}^{m,n}$ and L', L be positive integers with $L' < L$.

(a) The map

$$\widehat{}_{[0,L]} : C^\infty(\epsilon_{m,n[L]}(U), \mathbb{R}) \to \left\{ \text{functions of } U \text{ into } \mathbb{R}_{S[L]} \right\} \tag{4.41}$$

is defined by

$$\widehat{f}_{[0,L]}(x;\xi)$$

$$= \sum_{i_1=0,\ldots,i_m=0}^{L} \frac{1}{i_1! \ldots i_m!} \partial_1^{i_1} \ldots \partial_m^{i_m} f(\epsilon_{m,n[L]}(x)) \times s(x^1)^{i_1} \ldots s(x^m)^{i_m}.$$

$$\tag{4.42}$$

(b) The map

$$\widehat{}_{[L',L]} : C^\infty(\epsilon_{m,nL}(U), \mathbb{R}_{S[L']}) \to \left\{ \text{functions of } U \text{ into } \mathbb{R}_{S[L]} \right\} \tag{4.43}$$

is defined by

$$\widehat{f}_{[L',L]}(x;\xi) = \sum_{i_1=0,\ldots,i_m=0}^{L} \frac{1}{i_1! \ldots i_m!} \iota_{[L',L]}(\partial_1^{i_1} \ldots \partial_m^{i_m} f(\epsilon_{m,nL}(x)))$$

$$\times s(x^1)^{i_1} \ldots s(x^m)^{i_m}. \tag{4.44}$$

Equipped with these definitions the new class of functions can be defined. Briefly, GH^∞ functions have coefficient functions which are $\mathbb{R}_{S[L']}$-valued.

This class is sufficiently restricted to avoid ambiguous odd derivatives but wide enough to include constant Grassmann functions and supersymmetry transformations.

Definition 4.6.2   Suppose that U is open in $\mathbb{R}^{m,n}_{S[L]}$, with $L > 2n$. Let $L' = [\frac{1}{2}L]$, the greatest integer not greater than $\frac{1}{2}L$. Then $f : U \to \mathbb{R}_{S[L]}$ is said to be GH^∞ if and only if there exist, for each multiindex μ in M_n, a C^∞ function $f_{\underline{\mu}} : \epsilon_{m,nL}(U) \to \mathbb{R}_{S[L']}$ such that

$$f(x;\xi) = \sum_{\underline{\mu} \in M_n} \widehat{f_{\underline{\mu}}}_{[L',L]}(x)\xi^{\underline{\mu}} \tag{4.45}$$

for all $(x;\xi)$ in U. The set of all such functions is denoted $GH^\infty(U)$.

It will emerge in the following chapter, on supermanifolds, that the structures do not depend heavily on the particular value of L. Many of the properties possessed by G^∞ functions on $\mathbb{R}^{m,n}_S$ have their analogues for H^∞ and GH^∞ functions. Both $GH^\infty(U)$ and $H^\infty(U)$ are super algebras, but they are distinguished from $G^\infty(U)$ in that they are not modules over the full Grassmann algebra involved; $H^\infty(U)$ is not a module over any Grassmann algebra, while $GH^\infty(U)$ is only a module over the smaller algebra $\mathbb{R}_{S[L']}$. Further details may be found in [122].

4.7   The inverse function theorem

A complete analogue of the standard inverse function theorem holds for supersmooth functions; as a consequence, the implicit function theorem is also valid, which has important implications for the theory of sub-supermanifolds.

Theorem 4.7.1   *Suppose that U is an open subset of $\mathbb{R}^{m,n}_S$ (in the DeWitt topology) and that $F : U \to \mathbb{R}^{m,n}_S$ is G^∞. Then, if the $(m,n) \times (m,n)$ super matrix of derivatives $(\partial^S_k F^j)$ is invertible at a point $(x;\xi)$ in U, there exists a neighbourhood V of $(x;\xi)$ such that $F|_V : V \to F(V)$ has G^∞ inverse G; also the derivatives of G satisfy*

$$\sum_{k=1}^{m+n} \partial^S_j(F^k(p))\partial^S_k(G^\ell(F(p))) = \delta^\ell_j \text{ for } p \in V$$

$$\text{and} \quad \sum_{k=1}^{m+n} \partial^S_j(G^k(q))\partial^S_k(F^\ell(G(q))) = \delta^\ell_j \text{ for } q \in F(V). \tag{4.46}$$

Outline of proof Let F have components $(f^1, \ldots, f^m; \phi^1, \ldots, \phi^n)$. Since the matrix $(\partial_k^S F^j)$ is invertible at $(x; \xi)$, the $m \times m$ matrix $(\epsilon(\partial_i^E f^j))$ is invertible at $\epsilon_{m,n}(x; \xi)$, so that there exists an open subset V of U containing $(x; \xi)$ such that $F_{[\emptyset]}|_{V_{[\emptyset]}} : V_{[\emptyset]} \to F_{[\emptyset]}(V_{[\emptyset]})$ is invertible with C^∞ inverse. Let $G_{[\emptyset]}$ denote this inverse.

The theorem will now be proved by constructing a solution G to the equation

$$G \circ F(x; \xi) = (x; \xi) \tag{4.47}$$

by induction over the number of generators in the Grassmann algebra, and order by order in ξ.

Suppose that, for $i = 1, \ldots, m$,

$$f^i(x; \xi) = \sum_{\underline{\mu} \in M_n} f^i_{\underline{\mu}}(x) \xi^{\underline{\mu}}$$

$$= \sum_{\underline{\mu} \in M_n} \sum_{\underline{\lambda} \in M_\infty} \beta_{[\underline{\lambda}]} f^i_{\underline{\lambda}\,\underline{\mu}}(x) \xi^{\underline{\mu}}. \tag{4.48}$$

and for $k = 1, \ldots, n$,

$$\phi^k(x; \xi) = \sum_{\underline{\mu} \in M_n} \phi^k_{\underline{\mu}}(x) \xi^{\underline{\mu}}$$

$$= \sum_{\underline{\mu} \in M_n} \sum_{\underline{\lambda} \in M_\infty} \beta_{[\underline{\lambda}]} \phi^k_{\underline{\lambda}\,\underline{\mu}}(x) \xi^{\underline{\mu}}. \tag{4.49}$$

Functions $g^i_{\underline{\lambda}\mu}$, $\gamma^k_{\underline{\lambda}\,\mu}$ will be constructed such that

$$g^i(y; \phi) = \sum_{\underline{\mu} \in M_n} \sum_{\underline{\lambda} \in M_\infty} \beta_{[\underline{\lambda}]} g^i_{\underline{\lambda}\,\mu}(y) \phi^{\underline{\mu}}$$

$$\gamma^k(y; \phi) = \sum_{\underline{\mu} \in M_n} \sum_{\underline{\lambda} \in M_\infty} \beta_{[\underline{\lambda}]} \gamma^k_{\underline{\lambda}\,\mu}(y) \phi^{\underline{\mu}} \tag{4.50}$$

are the components of a function $G = (g; \gamma)$ which satisfies equation (4.47). A coefficient function such as $g_{\underline{\lambda}\mu}$ will be said to be of level (m, n) if $\underline{\lambda}$ is of length m and μ of length n.

At level $(0, 0)$, $g_{\underline{\emptyset}\,\underline{\emptyset}}$ must satisfy $g_{\underline{\emptyset}\,\underline{\emptyset}} \circ f_{[\emptyset]}|_{V_{[\emptyset]}} = \mathrm{id}_{V_{[\emptyset]}}$. This is achieved by setting $g_{\underline{\emptyset}\,\underline{\emptyset}} = (f_{\underline{\emptyset}\,\underline{\emptyset}}|_{V_{[\emptyset]}})^{-1}$, recalling that V has been chosen so that such an inverse exists.

At level $(0,1)$, $\gamma^k_{\underline{\emptyset}\mu}, k = 1, \ldots, n$ must satisfy

$$\xi^k = \gamma^k_{\underline{\emptyset}\mu}(f_{\underline{\emptyset}\,\underline{\emptyset}}(x))\phi^\mu_{\underline{\emptyset}\nu}(x)\xi^\nu \tag{4.51}$$

Since (by hypothesis) $\phi^\mu_{\underline{\emptyset}\nu}$ must be invertible, a unique $\gamma^k_{\mu\underline{\emptyset}}$ is determined. At level $(0,2)$ it is necessary that

$$0 = \partial^E_j g_{[\emptyset]}{}^i(f_{\underline{\emptyset}\,\underline{\emptyset}}(x))f^j_{\underline{\emptyset}\,\nu_1\nu_2}(x) + g^i_{\underline{\emptyset}\,\mu_1\mu_2}(f_{\underline{\emptyset}\,\underline{\emptyset}}(x))\phi^{\mu_1}_{\underline{\emptyset}\nu_1}\phi^{\mu_2}_{\underline{\emptyset}\nu_2} \tag{4.52}$$

which (again using the invertibility of $\phi^\mu_{\underline{\emptyset}\nu}$) determines $g_{\underline{\emptyset}\mu_1\mu_2}$ uniquely. At level $(1,0)$, $\gamma^\kappa_{\lambda\underline{\emptyset}}$ is determined by the requirement that

$$0 = \gamma^k_{\lambda\underline{\emptyset}}(f_{\underline{\emptyset}\,\underline{\emptyset}}(x)) + \gamma^k_{\underline{\emptyset}\mu}(f_{\underline{\emptyset}\,\underline{\emptyset}}(x))\phi^\mu_{\lambda\,\underline{\emptyset}}(x). \tag{4.53}$$

In a similar manner it can be shown inductively that all coefficient functions $g^i_{\underline{\lambda}\,\mu}$ and $\gamma^k_{\underline{\lambda}\,\mu}$ are uniquely determined by the requirement that $G \circ F(x;\xi) = (x;\xi)$. The rest of the theorem then immediately follows. ∎

4.8 Partitions of unity

In this section it is shown that G^∞ partitions of unity exist on any open subset of $\mathbb{R}^{m,n}_S$. Here, as always except when explicitly stated to the contrary, the DeWitt topology is used. The section is very brief, but provides a good example of the power of the DeWitt topology.

Theorem 4.8.1 *Suppose that U is open in $\mathbb{R}^{m,n}_S$. Let $\{U_\alpha | \alpha \in \Lambda\}$ be a locally finite open cover of U. Then there exist G^∞ functions $\{f_\alpha | \alpha \in \Lambda\}$ with the support of each f_α contained in U_α such that*

$$\sum_{\alpha \in \Lambda} f_\alpha = 1. \tag{4.54}$$

A collection of functions with these properties is said to be a partition of unity *on U subordinate to $\{U_\alpha | \alpha \in \Lambda\}$.*

Proof Suppose that $U = (\epsilon_{m,n})^{-1}(V)$ where V is open in $\mathbb{R}^m$, and for each α in Λ $U_\alpha = (\epsilon_{m,n})^{-1}(V_\alpha)$ with V_α open in $\mathbb{R}^m$. Then $\{V_\alpha | \alpha \in \Lambda\}$ is a locally finite open cover of V. Let $\{g_\alpha | \alpha \in \Lambda\}$ be a partition of unity on V subordinate to this open cover. Then, since the Grassmann analytic continuation of a constant function is simply a constant function taking the same value,

$$\sum_{\alpha \in \Lambda} \widehat{g}_\alpha = 1 \qquad (4.55)$$

using local finiteness and Theorem 4.2.3 which, as remarked before, extends to the infinite-dimensional case. Hence by taking $f_\alpha = \widehat{g}_\alpha$ the required partition of unity is obtained.

4.9 Superholomorphic functions of complex Grassmann variables

A very similar construction to that of Section 4.3 allows the definition of a superholomorphic function of a complex Grassmann algebra.

Definition 4.9.1   Let U be open in $\mathbb{C}_S^{m,n}$. Then $f : U \to \mathbb{R}_S$ is said to be *superholomorphic* or GC^ω on U if and only if there exists a collection $\left\{ f_{\underline{\mu}} | \underline{\mu} \in M_n \right\}$ of $\mathbb{C}_S$-valued functions which are holomorphic on $\epsilon_{m,n}(U)$ such that

$$f(z, \zeta) = \sum_{\underline{\mu} \in M_n} \widehat{f_{\underline{\mu}}}(z) \zeta^{\underline{\mu}} \qquad (4.56)$$

for each (z, ζ) in U. (As before this expansion is called the Grassmann analytic expansion of f and the functions $f_{\underline{\mu}}$ are called the Grassmann analytic coefficients of f.) The set of all such functions on U is denoted $GC^\omega(U)$.

There is also an analogue of H^∞ in the complex analytic case; this consist of functions of the form (4.56) with $f_{\underline{\mu}}$ taking values in $\mathbb{C}$. The set of such functions is denoted $HC^\omega(U)$. The properties of Grassmann analytic continuation mean that many features of complex analysis transfer simply to the super setting. Some of these will be discussed further in Chapter 14 which concerns super Riemann surfaces.

Chapter 5

Supermanifolds: The concrete approach

There exist in the literature a confusing number of different formulations of the concept of a supermanifold; these different approaches conceal a broad similarity of content, and it is the purpose of this chapter both to set up a useful working definition of a supermanifold, the G^∞ DeWitt supermanifolds of Section 5.1, and to begin to relate this to other definitions. This chapter also considers maps between supermanifolds.

Apart from differences of detail, there are two completely different ways of looking at supermanifolds: on the one hand one can, as in this chapter, take the concrete approach and define a supermanifold as a set with certain extra structure, and then proceed with other structures very much as in conventional differential geometry. Alternatively one may take the algebro-geometric approach and generalise the algebra of smooth functions on a conventional manifold to include anticommuting elements in a carefully prescribed way [19, 95]. This approach is described in Chapter 7. The two approaches are broadly interchangeable, in that one can realise the abstract algebras of the graded manifold approach as algebras of H^∞ functions on carefully constructed supermanifolds. These connections are explored in more detail in Chapter 8. It is unnecessary to be dogmatic about which approach to use; certain problems lend themselves to one approach and certain to another, while an individual's mathematical background will also influence the ease of working in a given approach.

In the concrete approach, a supermanifold is a space locally modelled on a superspace; the key ingredients of a supermanifold are thus the superspace on which it is locally modelled, the topology used on this superspace and the class of functions from which the transition functions for changing coordinates on overlapping coordinate patches must be drawn. In Section 5.3 a general definition of a supermanifold, appropriate for a wide class of these possibilities, is given. Before this, in Section 5.1, the particular case of a supermanifold modelled on $\mathbb{R}_S^{m,n}$ with the DeWitt topology, and transition functions required to be G^∞ is described; this is the definition which will usually be used in the rest of the book, and will be the meaning implied by the unqualified term *supermanifold*. In section Section 5.2 it is shown how a topology is put on a supermanifold.

Although various definitions are possible there are many features of supermanifolds, such as the structure of their function spaces, which are independent of the details in almost all cases. There are places in the literature where supermanifolds are applied without any particular class of supermanifold being specified; usually this does not matter, as only generic features of supermanifolds are required; occasionally one kind of supermanifold is specified, when in actual fact another kind is clearly implied. In particular, it may be stated that the algebro-geometric approach is being used when in fact some features of the concrete approach are involved.

Many supermanifolds – including all DeWitt supermanifolds – have associated with them in a natural way a conventional manifold, obtained by dividing out the nilpotent part of the coordinates of each point in a manner that is globally well defined; this manifold is known as the body of the supermanifold, and the construction means that the body map $\epsilon_{m,n} : \mathbb{R}_S^{m,n} \to \mathbb{R}^m$ induces a well defined map of the supermanifold onto its body. This construction, which is the key to the relation of the concrete and algebro-geometric approaches, is described in Section 5.4. The final section briefly considers complex supermanifolds.

5.1 G^∞ DeWitt supermanifolds

The supermanifolds constructed in this section may be considered to be the standard supermanifolds used in most applications in this book. They are modelled on the space $\mathbb{R}_S^{m,n}$ set up in Chapter 2, using the DeWitt topology. Thus the basic idea is to mimic the standard definition of a smooth manifold, with $\mathbb{R}_S^{m,n}$ replacing $\mathbb{R}^m$ and G^∞ functions replacing C^∞ functions.

Definition 5.1.1   Let $\mathcal{M}$ be a set, and let m and n be positive integers.

(a) An (m,n)-G^∞ *chart* on $\mathcal{M}$ is a pair (V, ψ) where V is a subset of $\mathcal{M}$ and ψ is a bijective mapping of V onto an open subset of $\mathbb{R}_S^{m,n}$ (in the DeWitt topology).

(b) An (m, n)-G^∞ *atlas* on $\mathcal{M}$ is a collection of charts $\{(V_\alpha, \psi_\alpha) | \alpha \in \Lambda\}$ such that

  i. $\cup_{\alpha \in \Lambda} V_\alpha = \mathcal{M}$
  ii. for each α, β in Λ such that $V_\alpha \cap V_\beta \neq \emptyset$ the map

$$\psi_\alpha \circ \psi_\beta^{-1} : \psi_\beta(V_\alpha \cap V_\beta) \to \psi_\alpha(V_\alpha \cap V_\beta) \tag{5.1}$$

  is G^∞.

(c) An (m, n)-G^∞ atlas $\{(V_\alpha, \psi_\alpha) | \alpha \in \Lambda\}$ on $\mathcal{M}$ which is not contained in any other such atlas on $\mathcal{M}$ is called a *complete* (m, n)-G^∞ atlas on $\mathcal{M}$.
(d) An (m, n)-G^∞ *De Witt supermanifold* consists of a set $\mathcal{M}$ together with a complete (m, n)-G^∞ atlas on $\mathcal{M}$.

The maps ψ_α are called coordinate maps and the sets V_α are called coordinate neighbourhoods.

It is useful to note that, as with a C^∞ atlas in conventional differential geometry [92], a (m, n)-G^∞ atlas can be extended to a complete atlas in a unique way. Thus to specify a supermanifold, it is sufficient to specify a set $\mathcal{M}$ and an (m, n)-G^∞ atlas on $\mathcal{M}$.

Some examples of G^∞ DeWitt supermanifolds will now be given, beginning with two standard, if trivial, examples which show that $\mathbb{R}_S^{m,n}$ and any open subset V of $\mathbb{R}_S^{m,n}$ (in the DeWitt topology) can be given the structure of an (m, n)-G^∞ DeWitt supermanifold in a natural way.

Example 5.1.2 Let $\mathcal{M} = \mathbb{R}_S^{m,n}$. Then $(\mathcal{M}, \mathrm{id})$ is a (m, n)-G^∞ chart on $\mathcal{M}$, and thus $\{(\mathcal{M}, \mathrm{id})\}$ is an (m, n)-G^∞ atlas on $\mathcal{M}$.

Example 5.1.3 Let V be an open subset of $\mathbb{R}_S^{m,n}$ (in the DeWitt topology). Then (V, ι) (where ι denotes inclusion) is an (m, n)-G^∞ chart on V and $\{(V, \iota)\}$ is an (m, n)-G^∞ atlas on V.

The next example gives the construction of real super projective space $\mathbb{SRP}^{m,n}$. This is defined by putting an equivalence relation on the subset U of $\mathbb{R}_S^{m+1,n}$ whose body is non-zero.

Example 5.1.4 Let $U \subset \mathbb{R}_S^{m+1,n}$ be the set $(\epsilon_{m+1,n})^{-1}(\mathbb{R}^{m+1} - \{0\})$. (The map $\epsilon_{m+1,n} : \mathbb{R}_S^{m+1,n} \to \mathbb{R}^{m+1}$ is defined in Equation (3.8).) Also let $\sim$ be the equivalence relation on U with $(x; \xi) \sim (x'; \xi')$ if and only if there exists an invertible even element ℓ of $\mathbb{R}_S$ such that

$$\begin{aligned} x^i &= \ell x'^i & i = 1, \ldots, m \\ \xi^j &= \ell \xi'^j & j = 1, \ldots, n. \end{aligned} \tag{5.2}$$

Then $\mathbb{SRP}^{m,n} = U/\sim$ may be given the structure of an (m,n)-G^∞ DeWitt supermanifold by defining the following atlas. For $i = 1, \ldots, m+1$ let

$$V_i = \left\{ [(x;\xi)] \mid (x;\xi) \in U, \epsilon(x^i) \neq 0 \right\}. \tag{5.3}$$

(Here $[(x;\xi)]$ denotes the equivalence class containing the point $(x;\xi)$.) Also define

$$\psi_i : V_i \to \mathbb{R}_S^{m,n}$$

$$[(x;\xi)] \mapsto \left(\frac{x^1}{x^i}, \ldots, \frac{\hat{x^i}}{x^i}, \ldots, \frac{x^{m+1}}{x^i}, \frac{\xi^1}{x^i}, \ldots, \frac{\xi^n}{x^i} \right) \tag{5.4}$$

(where the caret $\hat{\ }$ denotes omission of the argument). The maps ψ_i are well defined, since if $(x;\xi) \sim (x';\xi')$ and $\epsilon(x^i) \neq 0$, then

$$\frac{x^j}{x^i} = \frac{x'^j}{x'^i} \quad \text{for} \quad j = 1, \ldots, m+1 \quad \text{and} \quad \frac{\xi^j}{x^i} = \frac{\xi'^j}{x'^i} \quad \text{for} \quad j = 1, \ldots, n. \tag{5.5}$$

Since the transition functions are clearly G^∞, the space $\mathbb{SRP}^{m,n}$ has been given the structure of a (m,n)-G^∞ DeWitt supermanifold.

While this construction of projective superspace is the natural analogue of the standard construction of $\mathbb{RP}^m$, many of the standard constructions of quotient and algebraic manifolds do not produce DeWitt supermanifolds. For instance there is no nontrivial action of the additive group $\mathbb{Z}$ of integers on $\mathbb{R}_S^{0,1}$ which acts properly discontinuously with respect to the DeWitt topology, and so one cannot curl up the odd directions (or indeed the nilpotent parts of the even directions) to obtain tori as one might expect. Such constructions are possible when the finer topology is used, as is shown in Example 5.3.2. Also subsets of $\mathbb{R}_S^{m,n}$ which obey quite simple algebraic equations may fail to be supermanifolds. For instance the subset V of $\mathbb{R}_S^{0,2}$ consisting of points $(\xi^1;\xi^2)$ such that $\xi^1\xi^2 = 0$ is not a supermanifold under any definition. This is because of the algebraic properties of odd Grassmann elements. Further examples of supermanifolds are given in Chapter 8, where it is shown that all (m,n)-G^∞ DeWitt supermanifolds may be constructed from a vector bundle over a standard manifold. Other examples will occur in later chapters.

5.2 The topology of supermanifolds

In the previous section a G^∞ DeWitt supermanifold $\mathcal{M}$ with complete atlas $\{(V_\alpha, \psi_\alpha)|\alpha \in \Lambda\}$ was defined as a set with no a priori topology. However it is easy to use the supermanifold structure to define a topology on $\mathcal{M}$. The idea is essentially to define the topology by requiring that each coordinate map is a homeomorphism of the corresponding coordinate neighbourhood onto its image. The following theorem shows how this may be done in a unique way.

Theorem 5.2.1 *Let $\mathcal{M}$ be a (m, n)-G^∞ DeWitt supermanifold with complete atlas $\{(V_\alpha, \psi_\alpha)|\alpha \in \Lambda\}$. Let Γ_{DeWitt} be the collection of subsets of $\mathcal{M}$ consisting of sets $U \subset \mathcal{M}$ such that, for all $\alpha \in \Lambda$, $\psi_\alpha(U \cap V_\alpha)$ is open in $\mathbb{R}_S^{m,n}$ with the DeWitt topology. Then Γ_{DeWitt} is a topology on $\mathcal{M}$. (This topology will be referred to as the DeWitt topology of the supermanifold $\mathcal{M}$.)*

The proof of this theorem is exactly as for classical manifolds. Because the DeWitt topology of $\mathbb{R}_S^{m,n}$ is not Hausdorff, the DeWitt topology of a supermanifold is also non-Hausdorff.

Now in fact it is possible to define other topologies on $\mathbb{R}_S^{m,n}$, as described in Chapter 3. In each case, these are finer topologies than the DeWitt topology, and so allow a corresponding topology on $\mathcal{M}$, so that the coordinate maps are homeomorphisms of the coordinate neighbourhoods onto their images with respect to the given topology on $\mathbb{R}_S^{m,n}$. A generic construction is given in the following theorem.

Theorem 5.2.2 *Suppose that $\mathcal{T}$ is a topology on $\mathbb{R}_S^{m,n}$ which is finer than the DeWitt topology. Let $\mathcal{M}$ be a (m, n)-G^∞ DeWitt supermanifold with complete atlas $\{(V_\alpha, \psi_\alpha)|\alpha \in \Lambda\}$. Let $\Gamma_{\mathcal{T}}$ be the collection of subsets of $\mathcal{M}$ consisting of sets $U \subset \mathcal{M}$ such that for all $\alpha \in \Lambda$ $\psi_\alpha(U \cap V_\alpha)$ is open in $\mathbb{R}_S^{m,n}$ with the topology $\mathcal{T}$. Then $\Gamma_{\mathcal{T}}$ is a topology on $\mathcal{M}$. (This topology will be referred to as the $\mathcal{T}$ topology of the supermanifold $\mathcal{M}$.)*

Again the proof of this theorem is exactly as in the classical case.

It is occasionally useful to observe that a smooth supermanifold modelled on $\mathbb{R}_{S[L]}^{m,n}$ (in the manner described in the following section) can be given the structure of a $2^{L-1}(m+n)$-dimensional C^∞ manifold. This manifold will be referred to as the *fine structure* manifold of $\mathcal{M}$.

5.3   More general supermanifolds

In the literature there are a large number of different definitions of super-manifold. The purpose of this section is to describe many of these definitions, and to relate them to one another. The closely related algebro-geometric approach to supermanifolds, where algebras (generalising the function algebras on a smooth manifold) are the primary objects, are described in Chapter 7, while in Chapter 8 the relationship between the concrete and algebro-geometric approach is described. These more general structures are largely included for completeness; however supermanifolds modelled on finite-dimensional Grassmann algebras also provide useful stepping stones in some proofs relating to the standard DeWitt supermanifolds of section Section 5.1.

The key Definition 5.3.1 defines a generic supermanifold; the essential ingredients are the superspace, the topology on the superspace, and the class of allowed transition functions on overlapping coordinate systems; the first part of Definition 5.3.1 systemises these features, while the second part specifies the properties of the atlas required to give a set the corresponding supermanifold structure.

Definition 5.3.1

(a) Let $\mathbb{A}^{m,n}$ be an (m,n)-dimensional superspace. Let $\mathcal{T}$ be a topology on $\mathbb{A}^{m,n}$. Also, following [92], suppose that $\mathcal{F}$ is a collection of maps f with the following properties.

 (i) Each f in $\mathcal{F}$ is a homeomorphism whose domain and range are both open sets in $\mathbb{A}^{m,n}$.

 (ii) If a map f is in $\mathcal{F}$, then the restriction of f to any open subset of its domain is also in $\mathcal{F}$.

 (iii) If f is a mapping with domain $U = \cup_{i\in I}U_i$, and the restriction of f to each $U_i, i \in I$ is in $\mathcal{F}$, then f is in $\mathcal{F}$.

 (iv) If U is open in $\mathbb{A}^{m,n}$, then id_U, the identity map on U, is in $\mathcal{F}$.

 (v) If f is in $\mathcal{F}$ then its inverse f^{-1} is also in $\mathcal{F}$.

 (vi) If f, f' are in $\mathcal{F}$, and f has domain U, range V while f' has domain U' and range V', and $W = V \cap U'$ is not empty, then $f' \circ f|_{f^{-1}(W)}$ is in $\mathcal{F}$ with domain $f^{-1}(W)$ and range $f'(W)$.

Such a family $\mathcal{F}$ is said to be a *pseudogroup of transformations* of $(\mathbb{A}^{m,n}, \mathcal{T})$.

(b) Let $\mathcal{M}$ be a set

(i) An $(\mathbb{A}^{m,n}, \mathcal{T}, \mathcal{F})$-*chart* on $\mathcal{M}$ is a pair (V, ψ) where V is a subset of $\mathcal{M}$ and ψ is a bijective mapping of V onto an open subset of $\mathbb{A}^{m,n}$.

(ii) An $(\mathbb{A}^{m,n}, \mathcal{T}, \mathcal{F})$-*atlas* on $\mathcal{M}$ is a collection of charts $\{(V_\alpha, \psi_\alpha) | \alpha \in \Lambda\}$ such that

   A. $\cup_{\alpha \in \Lambda} V_\alpha = \mathcal{M}$

   B. for each α, β in Λ such that $V_\alpha \cap V_\beta \neq \emptyset$ the map

$$\psi_\alpha \circ \psi_\beta^{-1} : \psi_\beta(V_\alpha \cap V_\beta) \to \psi_\alpha(V_\alpha \cap V_\beta) \qquad (5.6)$$

   is in $\mathcal{F}$.

(iii) An $(\mathbb{A}^{m,n}, \mathcal{T}, \mathcal{F})$ atlas $\{(V_\alpha, \psi_\alpha) | \alpha \in \Lambda\}$ on $\mathcal{M}$ which is not contained in any other such atlas on $\mathcal{M}$ is called a *complete* $(\mathbb{A}^{m,n}, \mathcal{T}, \mathcal{F})$ atlas on $\mathcal{M}$.

(iv) An $(\mathbb{A}^{m,n}, \mathcal{T}, \mathcal{F})$ *supermanifold* consists of a set $\mathcal{M}$ together with a complete $(\mathbb{A}^{m,n}, \mathcal{T}, \mathcal{F})$ atlas on $\mathcal{M}$.

The maps ψ_α are called *coordinate maps* and the sets V_α are called *coordinate neighbourhoods*.

It may be checked that the collection G^∞ of maps f which are G^∞ bijective maps of between open subsets of $\mathbb{R}_S^{m,n}$ (with the DeWitt topology) onto another such set does form a pseudogroup of transformations of $\mathbb{R}_S^{m,n}$ (with the DeWitt topology). Thus the (m,n)-G^∞ DeWitt supermanifolds of section 2 are a special case of the definition given, specifically they are $(\mathbb{R}_S^{m,n}, \text{DeWitt}, G^\infty)$. Other possible classes of smooth supermanifold include $(\mathbb{R}_S^{m,n}, \mathcal{T}, G^\infty)$, $(\mathbb{R}_{S[L]}^{m,n}, \text{FDVS}, G^\infty)$, $(\mathbb{R}_{S[L]}^{m,n}, \text{ DeWitt}, G^\infty)$ (with $L > n$), $(\mathbb{R}_{S[L]}^{m,n}, \text{FDVS}, GH^\infty)$ (with $L > 2n$) and $(\mathbb{R}_{S[L]}^{m,n}, \text{ DeWitt }, H^\infty)$. (Here FDVS denotes the usual topology on a finite-dimensional vector space.) There are also complex supermanifolds, as described in Section 5.5. Examples of supermanifolds of this more general kind include those of Jadczyk and Pilch [84].

By using a finer topological vector space topology the class of possible supermanifolds increases; for instance, as the following examples show, it is possible to find non-trivial actions of discrete groups acting on $\mathbb{R}_S^{m,n}$ in such a way that the quotient space has a supermanifold structure which is not DeWitt, involving some compactification in odd or nilpotent directions.

Example 5.3.2 The space $S^1 \times S^1$ can be given the structure of a $(\mathbb{R}_{S[1]}^{1,1}, FDVS, G^\infty)$ supermanifold in the following way: regarding $S^1 \times S^1$ as $\mathbb{R}^2/R$, where R is the equivalence relation which identifies (x, y) and

(x', y') if an only if $x - x'$ and $y - y'$ are integers, let $\overline{(x, y)}$ denote the equivalence class containing the element (x, y) of $\mathbb{R}^2$. A collection of charts $\{(U_\alpha, \psi_\alpha) | \alpha = 1, \ldots, 4\}$ on $S^1 \times S^1$ is defined by

$$U_1 = \left\{ \overline{(x, y)} | \tfrac{1}{5} < x < \tfrac{4}{5}, \tfrac{1}{5} < y < \tfrac{4}{5} \right\}$$

$$U_2 = \left\{ \overline{(x, y)} | \tfrac{1}{5} < x < \tfrac{4}{5}, -\tfrac{2}{5} < y < \tfrac{2}{5} \right\}$$

$$U_3 = \left\{ \overline{(x, y)} | -\tfrac{2}{5} < x < \tfrac{2}{5}, -\tfrac{2}{5} < y < \tfrac{2}{5} \right\}$$

$$U_4 = \left\{ \overline{(x, y)} | -\tfrac{2}{5} < x < \tfrac{2}{5}, \tfrac{1}{5} < y < \tfrac{4}{5} \right\} \qquad (5.7)$$

together with

$$\psi_\alpha : U_\alpha \to \mathbb{R}^{1,1}_{S[1]} \quad \alpha = 1, \ldots, 4,$$

$$\overline{(x, y)} \mapsto (x1, y\beta_{[1]}), \quad \alpha = 1, \ldots, 4. \qquad (5.8)$$

In each case $\psi_\alpha(U_\alpha)$ is open in $\mathbb{R}^{1,1}_{S[1]}$ with the finite-dimensional vector space (FDVS) topology, but not the coarser DeWitt topology. A typical transition function is

$$\psi_2 \circ \psi_1^{-1} : \quad \left\{ (x1, y\beta_{[1]}) | \tfrac{1}{5} < x < \tfrac{4}{5}, \tfrac{1}{5} < y < \tfrac{2}{5} \right\}$$

$$\cup \left\{ (x1, y\beta_{[1]}) | \tfrac{1}{5} < x < \tfrac{4}{5}, \tfrac{3}{5} < y < \tfrac{4}{5} \right\}$$

$$\to \left\{ (x1, y\beta_{[1]}) | \tfrac{1}{5} < x < \tfrac{4}{5}, \tfrac{1}{5} < y < \tfrac{2}{5} \right\}$$

$$\cup \left\{ (x1, y\beta_{[1]}) | \tfrac{1}{5} < x < \tfrac{4}{5}, -\tfrac{2}{5} < y < -\tfrac{1}{5} \right\}$$

where

$$(x; \xi) \mapsto (x; \xi) \quad \text{if } (x; \xi) \in \left\{ (x1, y\beta_{[1]}) | \tfrac{1}{5} < x < \tfrac{4}{5}, \tfrac{1}{5} < y < \tfrac{2}{5} \right\}$$

and $(x; \xi) \mapsto (x; \xi - \beta_{[1]})$ if $(x; \xi) \in \left\{ (x1, y\beta_{[1]}) | \tfrac{1}{5} < x < \tfrac{4}{5}, \tfrac{3}{5} < y < \tfrac{4}{5} \right\}$.

$$(5.9)$$

This transition function (and the other $\psi_\alpha \circ \psi_\beta^{-1}$) are clearly G^∞, so that the set of charts provides an atlas on $S^1 \times S^1$.

The following example is of a supermanifold which is compact, and has non-Abelian fundamental group. It is constructed as the quotient of a super Lie group (c.f. Chapter 9) very much as an Iwasawa manifold is constructed from $\mathbb{C}^3$ as in [30].

Example 5.3.3 First suppose $\mathbb{R}^{1,2}_{S[1]}$ is identified with the set G of 3×3 matrices of the form

$$\begin{pmatrix} 1 & x & \xi^1 \\ 0 & 1 & \xi^2 \\ 0 & 0 & 1 \end{pmatrix}.$$

Noting that G forms a non-Abelian group under multiplication, let D be the subgroup of G consisting of matrices where

$$x = m1, \quad \xi^1 = n_1\beta_{[1]} \quad \text{and} \quad \xi^2 = n_2\beta_{[1]} \tag{5.10}$$

with m, n_1 and n_2 all integers. Then the quotient space G/D of left cosets of D in G can be given the structure of a $(\mathbb{R}^{1,2}_{S[1]}, FDVS, G^\infty)$ supermanifold. Details may be found in [119].

Other examples of quotient spaces which are supermanifolds when a fine topology is used (but not with the DeWitt topology) occur in the theory of super Riemann surfaces, as is explained in Chapter 14. Further consideration of non-DeWitt supermanifolds may be found in the work of Crane and Rabin [35, 36].

At the other end of the spectrum, the class of $(\mathbb{R}^{m,n}_S, \text{DeWitt}, H^\infty)$ supermanifolds is more restricted than that of (m,n)-G^∞ DeWitt supermanifolds. However, as is shown in Chapter 8, using a simple extension of a result due to Batchelor [11], any $(\mathbb{R}^{m,n}_S, \text{DeWitt}, G^\infty)$ supermanifold may in fact be given the structure of an $(\mathbb{R}^{m,n}_S, \text{DeWitt}, H^\infty)$ supermanifold, although not in a canonical way.

5.4 The body of a supermanifold

Every DeWitt supermanifold $\mathcal{M}$ of dimension (m,n) can be seen to have an underlying m-dimensional conventional manifold $\mathcal{M}_{[\emptyset]}$ associated with it, known as the body of the supermanifold. Many non-DeWitt supermanifolds also have bodies. The body of a supermanifold plays an important role in the analysis of the structure of supermanifolds, and of the spaces of supersmooth functions one can define on a supermanifold and considerable use is made in the following chapters of the concept of the body of a supermanifold. The construction is given in the following theorem, the basic idea being due to DeWitt [43], with the first formal definition having been given by Batchelor [11].

Theorem 5.4.1   *Let $\mathcal{M}$ be a $(\mathbb{R}_S^{m,n}, \text{DeWitt}, G^\infty)$ supermanifold with atlas $\{(V_\alpha, \psi_\alpha) | \alpha \in \Lambda\}$. Then*

(a) the relation $\sim$ defined on $\mathcal{M}$ by $p \sim q$ if and only if there exists $\alpha \in \Lambda$ such that both p and q lie in V_α and

$$\epsilon_{m,n}(\psi_\alpha(p)) = \epsilon_{m,n}(\psi_\alpha(q)) \tag{5.11}$$

is an equivalence relation.

(b) The space $\mathcal{M}_{[\emptyset]} = \mathcal{M}/\sim$ has the structure of an m-dimensional C^∞ manifold with atlas $\{(V_{[\emptyset]\alpha}, \psi_{[\emptyset]\alpha}) | \alpha \in \Lambda\}$, where

$$V_{[\emptyset]\alpha} = \{[p] | p \in V_\alpha\}$$
$$\psi_{[\emptyset]\alpha} : V_{[\emptyset]\alpha} \to \mathbb{R}^m$$
$$[p] \mapsto \epsilon_{m,n} \circ \psi_\alpha(p). \tag{5.12}$$

(Here square brackets $[\,]$ denote equivalence classes in $\mathcal{M}$ under $\sim$.)

The manifold $\mathcal{M}/\sim$ is called the body *of $\mathcal{M}$ and denoted $\mathcal{M}_{[\emptyset]}$. The canonical projection of $\mathcal{M}$ onto $\mathcal{M}_{[\emptyset]}$ is denoted by ϵ.*

Proof

(a) That the relation $\sim$ is reflexive and symmetric follows directly from the definition. To show that it is transitive, suppose that $p \sim q$ and $q \sim r$. Then there must exist α and β in Λ such that p and q are in V_α, q and r are in V_β, $\epsilon_{m,n}(\psi_\alpha(p)) = \epsilon_{m,n}(\psi_\alpha(q))$ and $\epsilon_{m,n}(\psi_\beta(q)) = \epsilon_{m,n}(\psi_\beta(r))$. Now, since q lies in $V_\alpha \cap V_\beta$, $\psi_\beta(q)$ lies in $\psi_\beta(V_\alpha \cap V_\beta)$. However $\epsilon_{m,n}(\psi_\beta(q)) = \epsilon_{m,n}(\psi_\beta(r))$, and thus, since $\psi(V_\alpha \cap V_\beta)$ is open in $\mathbb{R}_S^{m,n}$ in the DeWitt topology, $\psi_\beta(q)$ must also lie in $\psi_\beta(V_\alpha \cap V_\beta)$. Thus (since ψ is injective) r must lie in V_α. Now, since $\psi_\alpha \circ \psi_\beta^{-1}$ is G^∞, $\epsilon_{m,n}(\psi_\alpha(q)) = \epsilon_{m,n}(\psi_\alpha \circ \psi_\beta^{-1} \circ \psi_\beta(q)) = \epsilon_{m,n}(\psi_\alpha \circ \psi_\beta^{-1} \circ \psi_\beta(r)) = \epsilon_{m,n}(\psi_\alpha(r))$. Thus $\sim$ is transitive.

(b) Clearly

$$\cup_{\alpha \in \Lambda} V_\alpha = \mathcal{M}/\sim . \tag{5.13}$$

Also it follows from the definition of $\sim$ that the maps $\psi_{[\emptyset]\alpha}$ are well defined and injective. The image of $\psi_{[\emptyset]\alpha}$ is $\epsilon_{m,n}(\psi_\alpha(V_\alpha))$, which is open in $\mathbb{R}^m$ since $\psi_\alpha(V_\alpha)$ is open in $\mathbb{R}_S^{m,n}$ with the DeWitt topology. The Grassmann analytic continuation of Theorem 4.2.4 shows that if f

is a G^∞ map of an open subset U of $\mathbb{R}_S^{m,n}$ into $\mathbb{R}_S^{m,n}$, then $\epsilon_{m,n} \circ |_{\epsilon_{m,n}(U)}$
is a C^∞ map of $\epsilon_{m,n}(U)$ into $\mathbb{R}^m$. Thus the map

$$\psi_{[\emptyset]\alpha} \circ \psi_{[\emptyset]\beta}^{-1} : \epsilon_{m,n}(\psi_\beta(V_\alpha \cap V_\beta)) \to \epsilon_{m,n}(\psi_\alpha(V_\alpha \cap V_\beta)) \qquad (5.14)$$

is C^∞. This completes the proof.

This theorem remains true for more general supermanifolds based on
the DeWitt topology. The proof makes use of the fact that if a point p lies
in an open subset U of the superspace, then so does any other point q with
the same body as p. This is true for the DeWitt topology, but not for finer
topologies.

Because of the restrictive nature of the DeWitt topology, certain aspects
of a supermanifold are determined by its body. For instance, a supermanifold is compact if its body is compact, and simply connected if its body
is simply connected, while the fundamental group of a supermanifold is
simply the fundamental group of its body.

5.5   Complex supermanifolds

The general formalism described in Section 5.3 includes complex supermanifolds, that is, supermanifolds modelled locally on $\mathbb{C}_S^{m,n}$. In the complex
case the particular class of supermanifold usually used are the analytic De-
Witt supermanifolds, or $(\mathbb{C}_S^{m,n}, \text{DeWitt}, GC^\omega)$ supermanifolds. Although
such supermanifolds are subsumed in the general definition Definition 5.3.1
they are sufficiently important for their definition to be given explicitly
here.

Definition 5.5.1   Let $\mathcal{M}$ be a set, and let m and n be positive integers.

(a) An (m, n) GC^ω *chart* on $\mathcal{M}$ is a pair (V, ψ) where V is a subset of $\mathcal{M}$
    and ψ is a bijective mapping of V onto an open subset of $\mathbb{C}_S^{m,n}$ (in the
    DeWitt topology).

(b) An (m, n) GC^ω *atlas* on $\mathcal{M}$ is a collection of charts $\{(V_\alpha, \psi_\alpha) | \alpha \in \Lambda\}$
    such that

    i. $\cup_{\alpha \in \Lambda} V_\alpha = \mathcal{M}$

    ii.   for each α, β in Λ such that $V_\alpha \cap V_\beta \neq \emptyset$ the map

$$\psi_\alpha \circ \psi_\beta^{-1} : \psi_\beta(V_\alpha \cap V_\beta) \to \psi_\alpha(V_\alpha \cap V_\beta) \qquad (5.15)$$

    is GC^ω.

(c) An (m,n) GC^ω atlas $\{(V_\alpha, \psi_\alpha) | \alpha \in \Lambda\}$ on $\mathcal{M}$ which is not contained in any other such atlas on $\mathcal{M}$ is called a *complete (m,n) GC^ω atlas* on $\mathcal{M}$.

(d) An (m,n) GC^ω *DeWitt supermanifold* consists of a set $\mathcal{M}$ together with a complete (m,n) GC^ω atlas on $\mathcal{M}$.

Of course $\mathbb{C}_S^{m,n}$ is itself such a supermanifold, as is any open subset of $\mathbb{C}_S^{m,n}$. Less trivial examples are provided by super complex projective spaces $\mathbb{SCP}^{m,ns}$, which are defined very much as $\mathbb{SRP}^{m,n}$ (c.f. Example 5.1.4), and by the super Grassmannians and flag manifolds. A particular class of $(1,1)$-dimensional complex supermanifold known as a *super Riemann surface* is described in Chapter 14.

Chapter 6

Functions and vector fields

This chapter lays the foundations of differential geometry on supermanifolds by introducing the concept of a supersmooth or G^∞ function on a supermanifold.

The algebraic structure of the set $G^\infty(U)$ of G^∞ functions on an open subset U of a supermanifold is considered; globally it is shown to take the form of a super $\mathbb{R}_S$-module, while locally (when U is a coordinate neighbourhood) further structure is found which relates directly to the notion of algebro-geometric supermanifold. Other spaces of functions, such as $H^\infty(U)$, are also considered.

Having considered G^∞ $\mathbb{R}_S$-valued functions of a supermanifold, the notion of a G^∞ mapping between supermanifolds is developed, and also the key concept of superdiffeomorphism. Odd and even curves are defined, as well as the notion of super curve, which is required to give useful integral curves of odd vector fields.

The next step is to define tangent vectors; as in the classical case, this may be done either algebraically or more geometrically, using curves. Vector fields are then introduced; these may be considered either as tangent vectors varying in a G^∞ manner from point to point, or as derivations of the module $G^\infty(U)$; it is shown that the set $\mathcal{D}(U)$ of vector fields on U has the structure of a super Lie module over $G^\infty(U)$; additionally, when U is a coordinate neighbourhood then $\mathcal{D}(U)$ is a free module, with super basis the coordinate derivatives.

In the final section induced maps of tangent vectors are defined. This allows a definition of integral curve of a vector field. In the case of even vector fields, one parameter groups of transformations exist which induce a vector field very much as in the classical case. For odd vector fields the situation is rather different, it will be essential in applications to Lie groups to use the $(1,1)$-parameter super curves whose existence is demonstrated in Theorem 6.5.9. This difference arises because it is possible for the super commutator of an odd vector field with itself to be non-zero.

It is useful to establish notation for the chapter from the outset: $\mathcal{M}$ will denote a $(\mathbb{R}_S^{m,n}, \mathrm{DeWitt}, G^\infty)$ supermanifold with G^∞ structure $\{(V_\alpha, \psi_\alpha) | \alpha \in \Lambda\}$. For $i = 1, \ldots, m$, x_α^i denotes the ith even coordinate function ψ_α^i on V_α and, for $j = 0, \ldots, n$, ξ_α^j denotes the jth odd coordinate function ψ_α^{m+j}.

6.1　G^∞ functions on supermanifolds

An (m,n)-dimensional supermanifold is modelled locally on the space $\mathbb{R}_S^{m,n}$, and thus functions on supermanifolds inherit from this space a natural notion of superdifferentiability:

Definition 6.1.1　　Let U be an open subset of $\mathcal{M}$. Then

(a) a function $f : U \to \mathbb{R}_S$ is said to be G^∞ on U if, for each α in Λ such that $U \cap V_\alpha \neq \emptyset$, the function

$$f \circ \psi_\alpha^{-1} : \psi_\alpha(U \cap V_\alpha) \to \mathbb{R}_S \qquad (6.1)$$

is G^∞.

(b) The set of G^∞ functions on U is denoted $G^\infty(U)$.

(c) The function $f \circ \psi_\alpha^{-1} : \psi_\alpha(U \cap V_\alpha) \to \mathbb{R}_S$ is called the *local representative* of f in the coordinate neighbourhood V_α.

(d) If p is a point in $\mathcal{M}$, then $G^\infty(p)$ denotes the set of $\mathbb{R}_S$-valued functions which are G^∞ on some neighbourhood of p.

Because of the compatibility condition between coordinate maps ψ_α and ψ_β on overlapping coordinate neighbourhoods V_α and V_β, it is sufficient to check the condition (6.1) for every chart in some atlas on $\mathcal{M}$ compatible with the G^∞ structure on $\mathcal{M}$; the chain rule Theorem 4.4.2 then ensures that the condition must hold in every chart in the G^∞ structure.

A number of algebraic properties of the space of G^∞ functions on an open subset of $\mathbb{R}_S^{m,n}$ were established in Theorem 4.4.1; these lead directly to a similar set of algebraic properties for the set of functions $G^\infty(U)$, which are summarised in the following theorem.

Theorem 6.1.2　　*With the notation of Definition 6.1.1,*

(a) the set $G^\infty(U)$ is a super $\mathbb{R}_S$-module with the product of an element f of $G^\infty(U)$ by an element b of $\mathbb{R}_S$ defined by

$$(bf)(p) = b(f(p)) \tag{6.2}$$

for all p in U;
(b) the super $\mathbb{R}_S$-module $G^\infty(U)$ is also an algebra with product defined pointwise, that is,

$$fg(p) = f(p)g(p) \tag{6.3}$$

for all p in U.

Proof This theorem is proved by applying Theorem 4.4.1 to the local representatives of the functions in $G^\infty(U)$. ∎

On a coordinate neighbourhood U it is possible to give a more complete characterisation of the algebra $G^\infty(U)$. Suppose that f is in $G^\infty(U)$ and that ψ is the coordinate function on U; then $f \circ \psi^{-1}$ is G^∞ and so (by Definition 4.3.1) there exist, for $\underline{\mu} \in M_n$, functions $f_{\alpha\mu}$ in $C^\infty(\epsilon_{m,n}(\psi(U)), \mathbb{R}_S)$ such that

$$f \circ \psi^{-1}(y;\eta) = \sum_{\underline{\mu} \in M_n} \widehat{f_{\alpha\underline{\mu}}}(y)\eta^{\underline{\mu}}. \tag{6.4}$$

(Here ^ represents the Grassmann analytic continuation of Definition 4.2.2.) Thus

$$f = \sum_{\underline{\mu} \in M_n} f_{\underline{\mu}} \xi^{\underline{\mu}} \tag{6.5}$$

where

$$f_\mu = \widehat{f_{\alpha\underline{\mu}}} \circ \psi_\alpha, \tag{6.6}$$

which gives immediately the following proposition:

Proposition 6.1.3 *Suppose that U is a coordinate neighbourhood of the supermanifold $\mathcal{M}$. Then the super $\mathbb{R}_S$-modules $G^\infty(U)$ and $C^\infty(\epsilon_{m,n}(U), \mathbb{R}_S) \otimes \Lambda(\mathbb{R}^n)$ are isomorphic.*

Here the tensor product is simply a tensor product of an infinite-dimensional vector space with a finite-dimensional vector space, with no topology implied.

On other types of supermanifolds it will generally be a different class of function which plays the rôle of G^∞ functions on $(\mathbb{R}_S^{m,n}, \text{DeWitt}, G^\infty)$ supermanifolds. For instance, on an $(\mathbb{R}_S^{m,n}, \text{DeWitt}, H^\infty)$ supermanifold, there is a well-defined class of H^∞ functions defined via local representatives in a similar manner. The function rings $H^\infty(U)$ are not super $\mathbb{R}_S$-modules, but merely real vector spaces. As will emerge in Chapter 8, these function rings correspond to the algebras which arise in the algebro-geometric approach to supermanifolds.

A useful example of a collection of G^∞ functions is provided by the existence of partitions of unity.

Theorem 6.1.4 *Suppose that $\mathcal{M}$ is a supermanifold with compact body. Let $\{U_\alpha | \alpha \in \Gamma\}$ be an open cover of the supermanifold $\mathcal{M}$ such that the closure $\overline{U}_{\alpha[\emptyset]}$ of the body of each U_α is compact. Then there exist a functions $g_\alpha, \alpha \in \Gamma$ in $G^\infty(\mathcal{M})$ such that the support of each g_α is contained in U_α and has compact body, and also*

$$\sum_{\alpha \in \Gamma} g_\alpha = 1\,. \tag{6.7}$$

(The collection of functions $\{g_\alpha | \alpha \in \Gamma\}$ is called a partition of unity *subordinate to to the cover $\{U_\alpha | \alpha \in \Gamma\}$.)*

Proof First assume that each U_α is a coordinate neighbourhood with corresponding coordinate function ψ_α. Let $\{f_\alpha | \alpha \in \Gamma\}$ be a partition of unity on $\mathcal{M}_{[\emptyset]}$ subordinate to the cover $\{U_{\alpha[\emptyset]} | \alpha \in \Gamma\}$. (Such a partition of unity exists since $\mathcal{M}_{[\emptyset]}$ is paracompact.) Then functions g_α defined by

$$g_\alpha \circ \psi_\alpha^{-1}(y; \eta) = (f_\alpha \widehat{\circ \psi_{\alpha[\emptyset]}}^{-1})(y) \tag{6.8}$$

satisfy the conditions of the theorem. Since every open cover of $\mathcal{M}$ has a locally finite refinement by coordinate neighbourhoods, the theorem is true for any open cover. ∎

It will now be shown that any G^∞ function on an open subset can be extended (not in general uniquely) to a larger domain. The DeWitt topology allows such an extension theorem despite the considerable rigidity of functions in nilpotent directions.

Lemma 6.1.5 *Let V and U be open in $\mathcal{M}$ with $U \subset V$. Then, given f in $G^\infty(U)$, there exists a function $\tilde{f} : V \to \mathbb{R}_S$ which is G^∞ and satisfies*

$$\tilde{f}\Big|_U = f \tag{6.9}$$

Proof Suppose that $V = \bigcup_{\alpha \in \Lambda} U_\alpha$, where each U_α is a coordinate neighbourhood. Also let g_α be a partition of unity subordinate to the cover $\{U_\alpha\}$ of V. Then for each $\alpha \in \Lambda$ such that $U \cap U_\alpha$ is not empty there exist functions $f_{\alpha\underline{\mu}}, \underline{\mu} \in M_n$ in $C^\infty(U_{[\emptyset]} \cap U_{\alpha[\emptyset]})$ such that

$$f|_{U \cap U_\alpha} = \sum_{\underline{\mu} \in M_n} \widehat{f_{\alpha\underline{\mu}}} \xi_\alpha^{\underline{\mu}}. \tag{6.10}$$

Let $\tilde{f}_{\alpha\underline{\mu}}$ denote a C^∞ extension $f_{\alpha\underline{\mu}}$ to $U_{\alpha[\emptyset]}$. For those α in Λ where U_α does not intersect with U each $f_{\alpha\underline{\mu}}$ is defined to be zero. Then the function $\tilde{f} : V \to \mathbb{R}_S$ with

$$\tilde{f} = \sum_{\alpha \in \Lambda} \sum_{\underline{\mu} \in M_n} g_\alpha \widehat{\tilde{f}_{\alpha\underline{\mu}}} \xi_\alpha^{\underline{\mu}} \tag{6.11}$$

is in $G^\infty(V)$ and is equal to f when restricted to U.

6.2 Functions between supermanifolds

In the preceding section G^∞ functions of a supermanifold $\mathcal{M}$ into $\mathbb{R}_S$ have been defined. In this section the notion of a G^∞ function between supermanifolds is considered; before this the slightly simpler notion of G^∞ $\mathcal{M}$-valued functions on $\mathbb{R}_S^{r,s}$ is described; useful examples of such functions include super curves on a supermanifold.

Definition 6.2.1 Let U be an open subset of $\mathbb{R}_S^{r,s}$ where r and s are non-negative integers and let $f : U \to \mathcal{M}$. Then f is said to be G^∞ on U if for each α in Λ such that $f(U) \cap V_\alpha \neq \emptyset$ the function $\psi_\alpha \circ f : U \cap f^{-1}(V_\alpha) \to \mathbb{R}_S^{m,n}$ is G^∞.

This definition makes sense because $\psi_\alpha \circ f$ is a mapping of an open subset of $\mathbb{R}_S^{r,s}$ into $\mathbb{R}_S^{m,n}$, and thus Definition 4.3.2 can be applied. Particular examples of such maps are curves, which fall into three classes, even curves, odd curves and super curves. An even curve has domain contained in $\mathbb{R}_S^{1,0}$, an odd curve has domain $\mathbb{R}_S^{0,1}$ and a super curve has domain contained in $\mathbb{R}_S^{1,1}$.

Definition 6.2.2 Let A be an open interval in $\mathbb{R}$ containing 0.

(a) A G^∞ map c of $(\epsilon_{1,0})^{-1}(A)$ into $\mathcal{M}$ is said to be an *even curve* on $\mathcal{M}$. If p is a point in $\mathcal{M}$, the curve c is said to be *based at* p if $c(0) = p$.

(b) A G^∞ map γ of $(\epsilon_{0,1})^{-1}(A)$ into $\mathcal{M}$ is said to be a *odd curve* on $\mathcal{M}$. If p is a point in $\mathcal{M}$, the curve γ is said to be *based at* p if $\gamma(0) = p$.

(c) A G^∞ map of $(\epsilon_{1,1})^{-1}(A)$ into $\mathcal{M}$ is said to be a *super curve* on $\mathcal{M}$. If p is a point in $\mathcal{M}$, the super curve C on $\mathcal{M}$ is said to be *based at* p if $C(0;0) = p$.

It will be seen in the next section that curves and super curves will provide one approach to the tangent space of a supermanifold. The reason for considering super curves rather than simply curves depending on a single odd variable is related to the fact that the set of operators $\exp Q\tau$ with Q a fixed odd vector field and τ varying in $\mathbb{R}_{S1}$ does not form a group if $[Q, Q] \neq 0$; this is important when considering the exponential map of a super Lie group as in Section 9.6.

Suppose now that $\mathcal{N}$ is a $(\mathbb{R}_S^{r,s}, \mathrm{DeWitt}, G^\infty)$ supermanifold with atlas $\{(U_\beta, \phi_\beta), |\beta \in \Gamma\}$. Then it is possible to use the coordinate systems on $\mathcal{M}$ and $\mathcal{N}$ to define the notion of a G^∞ map between these two supermanifolds.

Definition 6.2.3   Suppose that $f : \mathcal{M} \to \mathcal{N}$. Then, if for all pairs α, β with $\alpha \in \Lambda$ and $\beta \in \Gamma$ such that $V_\alpha \cap f^{-1}(U_\beta) \neq \emptyset$ the function

$$\phi_\beta \circ \psi_\alpha^{-1} : \psi_\alpha(V_\alpha \cap f^{-1}(U_\beta)) \to \phi_\beta(U_\beta) \qquad (6.12)$$

is G^∞, then f is said to be G^∞.

A G^∞ map between supermanifolds is, if one takes a categorical point of view, the appropriate type of mapping to give a morphism between supermanifolds. In particular, it follows form the chain rule Theorem 4.4.2 that the composition of two G^∞ maps between supermanifolds is also G^∞. When there exists a G^∞ map between two supermanifolds which is bijective and has G^∞ inverse the two supermanifolds are equivalent, as is formalised in the concept of superdiffeomorphism:

Definition 6.2.4   If $f : \mathcal{M} \to \mathcal{N}$ is bijective, and both f and its inverse are G^∞, then f is said to be a *superdiffeomorphism* and $\mathcal{M}$ is said to be superdiffeomorphic to $\mathcal{N}$.

As one might expect, the relationship of superdiffeomorphism of supermanifolds is an equivalence relation, and classification of supermanifolds is carried out up to superdiffeomorphism. Also the set of superdiffeomorphisms of a supermanifold with itself forms a group under the composition of functions, known as the superdiffeomorphism group of $\mathcal{M}$, and abbreviated to $\mathrm{Sdiff}(\mathcal{M})$.

6.3 Tangent vectors

Tangent vectors at individual points in a supermanifold will now be defined, and then the concept of a vector field as a collection of tangent vectors will be developed. Two approaches to the concept of a tangent vector will be considered; one involves the action of a tangent vector on a function, and the other considers equivalence classes of super curves.

Definition 6.3.1 A *tangent vector* Y_p at a point p in a supermanifold $\mathcal{M}$ is a super $\mathbb{R}_S$-module morphism

$$Y_p : G^\infty(p) \to \mathbb{R}_S$$
$$f \mapsto Y_p f \qquad (6.13)$$

such that for all f, g in $G^\infty(p)$,

$$Y_p(fg) = Y_p f\, g(p) + (-1)^{|Y||f|} f(p) Y_p g. \qquad (6.14)$$

and for all constant functions B in $G^\infty(p)$

$$Y_p B = 0. \qquad (6.15)$$

The set of tangent vectors to $\mathcal{M}$ at p is denoted $T_p\mathcal{M}$.

The simplest example of a tangent vector are the coordinate derivatives $\left.\frac{\partial^E}{\partial x^i}\right|_p$ and $\left.\frac{\partial^O}{\partial \xi^j}\right|_p$, with respect to a specified set of coordinates $(x; \xi)$ about the point p. These will now be defined, and it will then be shown that the set $T_p\mathcal{M}$ of tangent vectors at p is a free super $\mathbb{R}_S$-module with the coordinate derivatives forming a super basis.

Definition 6.3.2 Let p be a point in $\mathcal{M}$, and let (V_α, ψ_α) be a chart about p, with corresponding local coordinate functions $(x; \xi)$. Then *coordinate derivatives* $\left.\frac{\partial^E}{\partial x^i}\right|_p$, $\left.\frac{\partial^O}{\partial \xi^j}\right|_p$ are defined by

$$\left.\frac{\partial^E}{\partial x^i}\right|_p : G^\infty(p) \to \mathbb{R}_S$$

$$\left.\frac{\partial^E}{\partial x^i}\right|_p f = \partial_i^E (f \circ \psi^{-1})(\psi(p)) \quad i = 1, \ldots, m$$

$$\left.\frac{\partial^O}{\partial \xi^j}\right|_p : G^\infty(p) \to \mathbb{R}_S$$

$$\left.\frac{\partial^O}{\partial \xi^j}\right|_p f = \partial_j^O (f \circ \psi^{-1})(\psi(p)) \quad j = 1, \ldots n. \qquad (6.16)$$

Also, where the Grassmann parity of the coordinate in question is unspecified, one may have

$$\left. \frac{\partial^S}{\partial X^k} \right|_p \; : \; G^\infty(p) \to \mathbb{R}_S$$

$$\left. \frac{\partial^S}{\partial X^k} \right|_p f = \partial_k^S (f \circ \psi^{-1})(\psi(p)) \quad k = 1, \ldots, m+n. \tag{6.17}$$

In the following proposition it is shown that each of these coordinate derivatives is a tangent vector.

Proposition 6.3.3   *Each of the even coordinate derivatives*

$$\left. \frac{\partial^E}{\partial x^i} \right|_p, i = 1, \ldots, m$$

is an even tangent vector at p, and each of the odd coordinate derivatives

$$\left. \frac{\partial^O}{\partial \xi^j} \right|_p, j = 0, \ldots, n$$

is an odd tangent vector at p.

Proof   The three necessary properties of the coordinate derivatives may immediately be deduced from Theorem 4.4.1, which establishes the linearity of odd and even derivatives, and also the super Leibniz property, for functions on $\mathbb{R}_S^{m,n}$.   ∎

The next theorem establishes the key feature of the set of tangent vectors at a point; in the case of conventional manifolds, the set of tangent vectors forms a vector space of the same dimension as the manifold. The corresponding result for a supermanifold is that the set of tangent vectors at a point form a free super $\mathbb{R}_S$-module of the same dimension as the supermanifold. The set $T_p\mathcal{M}$ is thus known as the tangent module.

Theorem 6.3.4   *The set $T_p\mathcal{M}$ of tangent vectors to $\mathcal{M}$ at p has the structure of a free (m,n)-dimensional super $\mathbb{R}_S$-module with multiplication by elements of $\mathbb{R}_S$ defined by*

$$(BY)_p(f) = B(Y_p f) \tag{6.18}$$

for all B in $\mathbb{R}_S$, Y in $T_p\mathcal{M}$ and f in $G^\infty(p)$, and addition defined by

$$(Y_p + Z_p)f = Y_p f + Z_p f. \tag{6.19}$$

Also the coordinate derivatives $\left\{ \frac{\partial^E}{\partial x^i} \Big|_p , \frac{\partial^O}{\partial \xi^j} \Big|_p , i = 1, \ldots, m, j = 0, \ldots, n \right\}$
form a super basis of this module, and, if Y_p is an element of $T_p\mathcal{M}$, then

$$Y_p = \sum_{i=1}^{m} Y_p x^i \left. \frac{\partial^E}{\partial x^i} \right|_p + \sum_{j=1}^{n} Y_p \xi^j \left. \frac{\partial^O}{\partial \xi^j} \right|_p . \qquad (6.20)$$

In order to prove this theorem, a Lemma is required, adapted from [72].

Lemma 6.3.5 *Let f be a function in $G^\infty(p)$. Then there exists a neighbourhood U of p contained in the domain of f and $m + n$ functions $f_k : U \to \mathbb{R}_S$, $k = 1, \ldots, m + n$, such that, for all $q \in U$,*

$$f(q) = f(p) + \sum_{i=1}^{m} (x^i(q) - x^i(p)) f_i(q) + \sum_{j=1}^{n} (\xi^j(q) - \xi^j(p)) f_{j+m}(q) \quad (6.21)$$

and

$$\left. \frac{\partial^E}{\partial x^i} \right|_p f = f_i(p) \quad \text{for } i = 1, \ldots, m$$

$$\left. \frac{\partial^O}{\partial \xi^j} \right|_p = f_{j+m}(p) \text{ for } j = 0, \ldots, n. \qquad (6.22)$$

Proof Let U be a coordinate neighbourhood of p which is contained in the domain of f, and let ψ be the corresponding coordinate function. Also, let $(a; \alpha) = \psi(p)$ and $(b; \beta) = \psi(q)$. Then

$f(q) - f(p)$

$= f \circ \psi^{-1}(b^1, \ldots, b^m; \beta^1, \ldots, \beta^n) - f \circ \psi^{-1}(a^1, b^2, \ldots, b^m; \beta^1, \ldots, \beta^n)$

$+ \cdots - \ldots$

$+ f \circ \psi^{-1}(a^1, \ldots, a^m; \alpha^1, \ldots, \alpha^{n-1}, \beta^n) - f \circ \psi^{-1}(a^1, \ldots, a^m; \alpha^1, \ldots, \alpha^n)$

$= \sum_{i=1}^{m} \int_0^1 (b^i - a^i)$

$\times \partial_i^E f(a^1, \ldots, a^{i-1}, a^i + t(b^i - a^i), b^{i+1}, \ldots, b^m; \beta^1, \ldots, \beta^n) dt$

$+ \sum_{j=1}^{n} \int_0^1 (\beta^\kappa - \alpha^\kappa)$

$\times \partial_\kappa^O f(a^1, \ldots, a^m; \alpha^1, \ldots, \alpha^{\kappa-1}, \alpha^\kappa + t(\beta^\kappa - \alpha^\kappa), \beta^{\kappa+1}, \ldots, \beta^n) dt$

$$(6.23)$$

Thus

$$f(q) - f(p) = \sum_{i=1}^{m}(x^i(q) - x^i(p))f_i(q) + \sum_{j=1}^{n}(\xi^j(q) - \xi^j(p))f_{j+m}(q) \quad (6.24)$$

where for $i = 1, \ldots, m$,

$$f_i(q) = \int_0^1 \partial_i^E f(a^1, \ldots, a^{i-1}, a^i + t(b^i - a^i), b^{i+1}, \ldots, b^m; \beta^1, \ldots, \beta^n) \mathrm{d}t$$

$$(6.25)$$

and for $j = 0, \ldots, n$,

$$f_{j+m}(q)$$

$$= \int_0^1 \partial_j^O f(a^1, \ldots, a^m; \alpha^1, \ldots, \alpha^{\kappa-1}, \alpha^\kappa + t(\beta^\kappa - \alpha^\kappa), \beta^{\kappa+1}, \ldots, \beta^n) \mathrm{d}t.$$

$$(6.26)$$

Thus $f(q)$ can be expressed in the required form, and explicit calculation shows that Equation (6.22) holds. ∎

Proof of Theorem 6.3.4 It follows immediately from the definition of addition and scalar multiplication for tangent vectors that $T_p\mathcal{M}$ is a super $\mathbb{R}_S$-module. Since

$$\left.\frac{\partial^S}{\partial X^k}\right|_p X^j = \delta_k^j \quad (6.27)$$

the tangent vectors $\left.\frac{\partial^S}{\partial X^k}\right|_p, k = 1, \ldots, m+n$ are linearly independent. Now suppose that f is a function in $G^\infty(p)$. Then, expressing f as in the previous lemma, and applying the Leibniz rule for tangent vectors,

$$Y_p f = \sum_{i=1}^{m} Y x^i \left.\frac{\partial^E}{\partial x^i}\right|_p f + \sum_{j=1}^{n} Y \xi^j \left.\frac{\partial^O}{\partial \xi^j}\right|_p f \quad (6.28)$$

which shows that $T_p\mathcal{M}$ is a free super $\mathbb{R}_S$-module of dimension (m, n) with the required super basis, and also that the coefficients in the expansion of Y_p in terms of this basis have the required values. ∎

An alternative approach to the tangent space is to consider equivalence classes of curves. This may be done by using even curves for even tangent vectors and odd curves for odd tangent vectors very much as in the classical setting. A further possibility, which is important when extending these ideas to vector fields, is to use super curves. Two equivalence relations, denoted $\sim_E$ and $\sim_O$, may be put on the set of super curves on $\mathcal{M}$ based at a point p. Letting $(t; \tau)$ denote local coordinates on $\mathbb{R}_S^{1,1}$, which contains the domain of any super curve, the first equivalence relation is defined by the condition

$$C_1 \sim_E C_2 \quad \text{if and only if}$$
$$\frac{\partial^E}{\partial t}(f \circ C_1)(0; 0) \;=\; \frac{\partial^E}{\partial t}(f \circ C_2)(0; 0) \text{ for all } f \in G^\infty(p). \quad (6.29)$$

If $[C]_E$ denotes the equivalence class of the curve C under $\sim_E$, one may regard $[C]_E$ as a map of $G^\infty(p)$ into $\mathbb{R}_S$ with

$$[C]_E(f) = \frac{\partial^E}{\partial t}(f \circ C)(0; 0). \quad (6.30)$$

It follows from the properties of the derivative $\frac{\partial^E}{\partial t}$ (Theorem 4.4.1) that $[C]_E$ defines an even tangent vector at p.

The second equivalence relation (which leads to odd tangent vectors) involves the super derivative $\mathcal{D}_T$ for G^∞ functions on $\mathbb{R}_S^{0,1}$, which is defined by

$$\mathcal{D}_T f(t, \tau) = \frac{\partial^E}{\partial t} f + \tau \frac{\partial^O}{\partial \tau} f. \quad (6.31)$$

It may easily verified that

$$\mathcal{D}_T^2 = \frac{\partial^E}{\partial t}, \quad (6.32)$$

which is useful when considering the semigroups generated by odd vector fields.

The equivalence relation $\sim_O$ is defined by

$$C_1 \sim_O C_2 \quad \text{if and only if}$$
$$\mathcal{D}_T(f \circ C_1)(0; 0) \;=\; \mathcal{D}_T(f \circ C_2)(0; 0) \quad \text{for all } f \in G^\infty(p). \quad (6.33)$$

In this case, if $[C]_O$ denotes the equivalence class of the curve C under $\sim_O$, one may regard $[C]_O$ as a map of $G^\infty(p)$ into $\mathbb{R}_S$ with

$$[C]_O(f) = \mathcal{D}_T(f \circ C)(0; 0). \quad (6.34)$$

One then finds that $[C]_O$ is an odd tangent vector.

Conversely, any even tangent vector Y_p with local coordinate expansion

$$Y_p = \sum_{i=1}^{m} Y^i \left. \frac{\partial^E}{\partial x^i} \right|_p + \sum_{j=1}^{n} Y^j \left. \frac{\partial^O}{\partial \xi^j} \right|_p \qquad (6.35)$$

is 'even tangent' to some curve C on $\mathcal{M}$ based at p with

$$\psi^k \circ C(t; \tau) = \psi^k(p) + t Y^k. \qquad (6.36)$$

Similarly an odd tangent vector Ξ_p is 'odd tangent' to the super curve C on $\mathcal{M}$ based at p with

$$\psi^k \circ C(t; \tau) = \psi^k(p) + \tau \Xi^k, \quad k = 1, \ldots, m+n. \qquad (6.37)$$

This establishes the following theorem:

Theorem 6.3.6 *The even part of the tangent space at p is in one-to-one correspondence with the $\sim_E$ equivalence classes of super curves on $\mathcal{M}$ based at p. The odd part of the tangent space at p is in one-to-one correspondence with the $\sim_O$ equivalence classes of super curves on $\mathcal{M}$ based at p.*

6.4 Vector fields

As with conventional manifolds, there are two equivalent approaches to the concept of vector field. One may either define a vector field on an open subset U of the supermanifold $\mathcal{M}$ to be a collection of tangent vectors, one for each point in U, varying smoothly as one moves about U, or as a super derivation of the super $\mathbb{R}_S$-module $G^\infty(U)$.

Definition 6.4.1 A *vector field* on an open subset U of the superman-ifold $\mathcal{M}$ is a collection Y of tangent vectors $\{Y_p | p \in U\}$ such that for each function f in $G^\infty(U)$ the function $Yf : U \to \mathbb{R}_S$ defined by

$$Yf(p) = Y_p f \qquad (6.38)$$

is also G^∞. The set of vector fields on U is denoted $\mathcal{D}(U)$.

It follows directly from this definition that a vector field Y defines a mapping of $G^\infty(U)$ into itself. Examples of vector fields are provided by the coordinate derivatives. A vector field Y on U is said to be even if Y_p is even for each point p in U, and odd if each Y_p is odd.

Lemma 6.4.2 *Suppose that U is a coordinate neighbourhood in $\mathcal{M}$, with corresponding coordinate function ψ. For $i = 1, \ldots, m$ let $\frac{\partial^E}{\partial x^i}$ denote the* *collection of tangent vectors* $\left\{ \frac{\partial^E}{\partial x^i} \Big|_p \,|\, p \in U \right\}$. *Also, for* $j = 0, \ldots, n$, *let*

$\frac{\partial^O}{\partial \xi^j}$ *denote the collection of tangent vectors* $\left\{ \frac{\partial^O}{\partial \xi^j} \Big|_p \,|\, p \in U \right\}$. *Then each*

$\frac{\partial^E}{\partial x^i}$ *and* $\frac{\partial^O}{\partial \xi^j}$ *is a vector field on* U.

Proof Let f be a function in $G^\infty(U)$. Then

$$\frac{\partial^E}{\partial x^i} f = \partial_i^E (f \circ \psi^{-1}) \circ \psi \tag{6.39}$$

which is G^∞ since $f \circ \psi^{-1}$ is G^∞. Thus $\frac{\partial^E}{\partial x^i}$ is a vector field on U. Similarly $\frac{\partial^O}{\partial \xi^j}$ is a vector field. ∎

Various properties of vector fields will now be established.

Theorem 6.4.3

(a) *$\mathcal{D}(U)$ is a super $G^\infty(U)$ module with multiplication of a vector field Y by a function g defined by*

$$(gY)_p = g(p)Y_p \,. \tag{6.40}$$

(b) *Any element Y of $\mathcal{D}(U)$ defines a super derivation of $G^\infty(U)$.*

(c) *$\mathcal{D}(U)$ is a super Lie $\mathbb{R}_S$-module, with bracket operation defined by*

$$[Z, Y] = ZY - (-1)^{|Z||Y|} YZ \tag{6.41}$$

for each pair of vector fields Z and Y.

(d) *If U is a coordinate neighbourhood then $\mathcal{D}(U)$ is a free super $G^\infty(U)$ module of dimension (m, n) and the coordinate derivatives $\frac{\partial^E}{\partial x^i}, i = 1, \ldots, m$, $\frac{\partial^O}{\partial \xi^j}, j = 0, \ldots, n$ form a super basis of $\mathcal{D}(U)$. Also, if $Y \in \mathcal{D}(U)$,*

$$Y = \sum_{i=1}^{m} Y x^i \frac{\partial^E}{\partial x^i} + \sum_{j=1}^{n} Y \xi^j \frac{\partial^O}{\partial \xi^j} \,. \tag{6.42}$$

Proof

(a) Suppose that f, g are in $G^\infty(U)$ and $Y \in \mathcal{D}(U)$. Then

$$(gY)f = g Y f \tag{6.43}$$

and thus (since $G^\infty(U)$ is closed under multiplication by Theorem 6.1.2) $(gY)f$ is G^∞ and thus gY is a vector field. Suppose that h is also in

$G^\infty(U)$. Then

$$
\begin{aligned}
(h(gY))_p &= h(p)(gY)_p \\
&= h(p)g(p)Y_p \\
&= hg(p)Y_p \\
&= ((hg)Y)_p.
\end{aligned}
\tag{6.44}
$$

Thus $\mathcal{D}(U)$ is a $G^\infty(U)$ module. Also,

$$
\begin{aligned}
|(gY)_p f| &= |g(p)| + |Y_p f|, \\
|(gY)f| &= |g| + |Yf|
\end{aligned}
\tag{6.45}
$$

and thus $\mathcal{D}(U)$ is a super $G^\infty(U)$ module.

(b) This result follows immediately from the Leibniz property of tangent vectors.

(c) It may be verified by explicit calculation that the super commutator $[Z, Y]$ of two vector fields Z and Y is also a vector field, with $|[Z, Y]| = |Z| + |Y|$. The super antisymmetry and Jacobi identities may also be verified by explicit calculation.

(d) This result follows from Theorem 6.3.4.

∎

The alternative approach to vector fields, which involves derivations, will now be considered. Letting $\mathrm{Der}(U)$ denote the set of super derivations of the super $\mathbb{R}_S$-module $G^\infty(U)$, it may be recalled from Proposition 2.4.4 that $\mathrm{Der}(U)$ is a super Lie $\mathbb{R}_S$-module. In order to establish that $\mathrm{Der}(U)$ and $\mathcal{D}(U)$ are isomorphic (both as super $G^\infty(U)$-modules and as super Lie $\mathbb{R}_S$-modules), it must first be shown that a derivation on $G^\infty(U)$ defines a unique derivation on $G^\infty(V)$ when V is an open subset of U. This is not trivial because a function in $G^\infty(V)$ does not have a unique extension to a functions in $G^\infty(U)$, although Lemma 6.1.5 ensures that an extension exists.

Lemma 6.4.4 *Suppose that P is in $\mathrm{Der}(U)$ and that V is an open subset of U. Then there is a derivation $P|_V$ in $\mathrm{Der}(V)$ (which is called the restriction of P to V) defined by*

$$
P|_V f = P\tilde{f}\Big|_V
\tag{6.46}
$$

where f is in $G^\infty(V)$ and $\tilde{f}$ is some G^∞ extension of f to U.

Proof It is sufficient to show that the action of $P|_V$ is independent of the choice of extension $\tilde{f}$ of f, since $P|_V$ will then acquire the necessary features of a derivation from P.

Suppose that $\tilde{f}_1$ and $\tilde{f}_2$ are two (possibly different) extensions of f to U. Also suppose that p is a point in V and g is an even function in $G^\infty(U)$ such that $g(p)$ is not zero but g is zero on the complement of V in U. (The existence of such a function can be deduced from the existence of partitions of unity, or demonstrated directly by arguments similar to those in the proof of Theorem 6.1.4.) Then, since P is a super derivation,

$$P(g(\tilde{f}_1 - \tilde{f}_2))(p) = P(g)(p)(\tilde{f}_1(p) - \tilde{f}_2(p)) + g(p)P(\tilde{f}_1 - \tilde{f}_2)(p). \quad (6.47)$$

Hence, since $g(\tilde{f}_1 - \tilde{f}_2)$ is identically zero, $\tilde{f}_1(p) = \tilde{f}_2(p)$ and $g(p)$ is not zero,

$$P(\tilde{f}_1 - \tilde{f}_2)(p) = 0. \quad (6.48)$$

Since p is an arbitrary point in V,

$$P(\tilde{f}_1 - \tilde{f}_2)\Big|_V = 0, \quad (6.49)$$

and so $P|_V$ is well defined. ∎

It will now be shown that the super $G^\infty(U)$-modules $\mathcal{D}(U)$ and $\mathrm{Der}(U)$ are isomorphic, and so a vector field can alternatively be defined as a super derivation.

Theorem 6.4.5 *The super $G^\infty(U)$-modules $\mathrm{Der}(U)$ and $\mathcal{D}(U)$ are isomorphic. They are also isomorphic as super Lie $\mathbb{R}_S$-modules.*

Proof Define the map $\iota : \mathcal{D}(U) \to \mathrm{Der}(U)$ to be inclusion. (This is possible since any Y in $\mathcal{D}(U)$ defines a super derivation of $G^\infty(U)$ by Theorem 6.4.3.) Then explicit calculation shows that ι defines a homomorphism of both super $G^\infty(U)$-modules and super Lie $\mathbb{R}_S$-modules. Also define $\kappa : \mathrm{Der}(U) \to \mathcal{D}(U)$ by setting

$$(\kappa(P))|_p F = P|_{V_F} F(p) \quad (6.50)$$

when F is a function in $G^\infty(p)$, and V_F is its domain. Then κ is the inverse of ι, and so ι must be an isomorphism. ∎

6.5 Induced maps and integral curves

Corresponding to any G^∞ map between supermanifolds there are induced maps of tangent spaces, exactly as on classical manifolds. For completeness, the appropriate definition will be given.

Definition 6.5.1 Suppose that f is a G^∞ map of a supermanifold $\mathcal{M}$ into a supermanifold $\mathcal{N}$, and that p is a point in $\mathcal{M}$. Then the induced map $f_* : T_p\mathcal{M} \to T_{f(q)}\mathcal{N}$ is defined by

$$f_*(Y_p)(g) = Y_p(g \circ f) \tag{6.51}$$

for every g in $G^\infty(f(p))$.

This definition allows the notion of an *immersion* as a mapping $f : \mathcal{M} \to \mathcal{N}$ which induces an injective map $f_* : T_p\mathcal{M} \to T_{f(p)}\mathcal{N}$ at every point p of $\mathcal{M}$. If f itself is injective then the immersion is said to be an *embedding*. An application of superembeddings in supersymmetric physics is described in Section 13.3.

Examples of induced maps occur in the two equivalence relations defined above on the space of super curves. If C_1 and C_2 are super curves on M, then $C_1 \sim_E C_2$ if and only if

$$C_{1*}\left(\frac{\partial^E}{\partial t}\right) = C_{2*}\left(\frac{\partial^E}{\partial t}\right) \tag{6.52}$$

and $C_1 \sim_O C_2$ if and only if

$$C_{1*}\left(\frac{\partial^E}{\partial t} + \tau\frac{\partial^O}{\partial \tau}\right) = C_{2*}\left(\frac{\partial^E}{\partial t} + \tau\frac{\partial^O}{\partial \tau}\right). \tag{6.53}$$

The notion of induced map leads naturally to the question of integral curves; analogues of the classical theory of integral curves on manifolds are obtained if super curves are used in the following manner. (At this stage super curves simply allow a unified picture. It will be seen below that they are required to give an analogue of one parameter groups of transformations corresponding to an odd vector field; also when super Lie groups are considered in Chapter 9 super curves will be necessary for consideration of the semigroup generated by an odd Lie algebra element.)

Definition 6.5.2 Let Y be an even vector field on a supermanifold $\mathcal{M}$ and p be a point in $\mathcal{M}$. Then the super curve $C : (\epsilon_{1,1})^{-1}(A) \to \mathcal{M}$ (with

A an open interval in the real line) is said to be an *integral curve* of Y based at p if

$$C_* \left(\frac{\partial^E}{\partial t} \bigg|_{t;\tau} \right) = Y_{C(t;\tau)} \tag{6.54}$$

for all $(t;\tau)$ in $(\epsilon_{1,1})^{-1}(A)$ and

$$C(0;0) = p. \tag{6.55}$$

The following theorem proves the existence of integral super curves of even vector fields; uniqueness however requires more than a simple initial condition, unless the super curves are restricted to be simply even.

Theorem 6.5.3 *Suppose that Y is an even vector field on $\mathcal{M}$, and Ξ_p is an odd tangent vector at a point p in $\mathcal{M}$. Then for any sufficiently small positive real number a there exists a unique integral super curve C of Y based at p with domain $(\epsilon_{1,1})^{-1}(I_a)$ (where I_a denotes the open interval $(-a,a)$) such that*

$$C_* \left(\frac{\partial^O}{\partial \tau} \bigg|_{(0;0)} \right) = \Xi_p. \tag{6.56}$$

Proof Let $(X) = (x;\xi)$ be coordinates about p, defined on the subset U of $\mathcal{M}$ containing p; then C is an integral super curve based at p for Y on U if an only if the components C^k of C in the coordinates (X) satisfy

$$\frac{\partial^E}{\partial t} C^k(t;\tau) = Y^k(C(t;\tau)),$$
$$\text{and} \quad C^k(0;0) = X^k(p), \qquad k = 1,\ldots,m+n. \tag{6.57}$$

Expanding $C^k(t;\tau)$ as

$$C^k(t;\tau) = A^k(t) + \tau B^k(t) \tag{6.58}$$

gives

$$\frac{\partial^E}{\partial t} A^k(t) = Y^k(A(t)) \tag{6.59}$$

$$\text{and} \quad \frac{\partial^E}{\partial t}(B^k(t)) = B^j(t)\partial_j^S Y^k(A(t)). \tag{6.60}$$

Also the initial conditions imply that $A(t)$ and $B(t)$ must satisfy the initial conditions

$$A^k(0) = X^k(p)$$
$$B^k(0) = \Xi_p(X^k). \qquad (6.61)$$

Restricting to the body of $(\epsilon_{1,1})^{-1}(I_a)$, for a sufficiently small these equations for $A(t), B(t)$ have a unique solution which can then be extended uniquely to all of $(\epsilon_{1,0})^{-1}(I_a)$, leading to a unique super curve with the required properties. (The vector $C_* \left(\left. \frac{\partial O}{\partial \tau} \right|_{(0;0)} \right)$ will be referred to as the odd tangent to C at p.) ∎

Corollary 6.5.4 *An even vector field Y on a supermanifold $\mathcal{M}$ has a unique even integral curve based at p with odd tangent at p equal to a given odd tangent vector Ξ_p at p.*

The corresponding notion for an odd super curve will now be described.

Definition 6.5.5 Let Y be an odd vector field on a supermanifold $\mathcal{M}$ and p be a point in $\mathcal{M}$. Then the super curve $C : (\epsilon_{1,1})^{-1}(A) \to \mathcal{M}$ (where A is an open interval in the real line containing 0) is said to be an *integral super curve* of Y based at p if

$$C_* \left(\left. \frac{\partial O}{\partial \tau} + \tau \frac{\partial E}{\partial t} \right|_{(t;\tau)} \right) = Y_{C(t;\tau)} \qquad (6.62)$$

for all $(t; \tau)$ in $(\epsilon_{1,1})^{-1}(A)$ and

$$C(0; 0) = p. \qquad (6.63)$$

The important result, that there exists a unique super curve based at a given point integral to a given odd vector field, will now be proved.

Theorem 6.5.6 *Let Ξ be an odd vector field on a supermanifold $\mathcal{M}$. Then for any sufficiently small positive real number a there exists a unique integral super curve C of Ξ based at p with domain $(\epsilon_{1,1})^{-1}(I_a)$.*

Proof Let $(X) = (x; \xi)$ be coordinates about p, defined on the subset U of $\mathcal{M}$; then C is an integral super curve based at p for Ξ on U if an only if the components C^k of C in the coordinates (X) satisfy

$$\left(\frac{\partial^O}{\partial \tau} + \tau \frac{\partial^E}{\partial t}\right) C^k(t;\tau) = \Xi^k(C(t;\tau)),$$

$$\text{and} \quad C^k(0;0) = X^k(p) \qquad k = 1,\ldots,m+n. \qquad (6.64)$$

Expanding $C^k(t;\tau)$ as

$$C^k(t;\tau) = A^k(t) + \tau B^k(t) \qquad (6.65)$$

gives

$$B^k(t) = \Xi^k(A(t)) \qquad (6.66)$$

$$\text{and} \quad \frac{\partial^E}{\partial t}(A^k(t)) = B^j(t)\partial_j^S \Xi^k(A(t)). \qquad (6.67)$$

Thus

$$\frac{\partial^E}{\partial t}(A^k(t)) = \Xi^j(A(t))\partial_j^S \Xi^k(A(t)). \qquad (6.68)$$

For sufficiently small a this equation for $A(t)$ has a unique solution on I_a given the initial condition, and so the integral curve is uniquely determined.

Remark 6.5.7 *The existence of even integral curves for even vector fields on a supermanifold means that an even vector field induces a local one parameter group of local transformations in a neighbourhood of any given point exactly as in the classical case.*

In more detail, with $I_a = (-a,a)$ as before, a local 1 parameter super group of local transformations on $I_a \times U$ is a mapping $\phi : I_a \times U \to \mathcal{M}$ such that

(a) For each $t \in I_a$ the mapping $\phi_t : p \mapsto \phi_t$ is a diffeomorphism of U onto its image.

(b) $\phi_t(\phi_s p) = \phi_{t+s}p$ for t, s and p such that $\phi_s p \in U$ and s, t and $s+t$ are in I_a.

If Y is an even vector field on the supermanifold $\mathcal{M}$ and p is a point in M, then there exists a neighbourhood U of p, a positive real number a and a local 1 parameter group of local transformations ϕ on $I_a \times U$ which induces Y in that at each q in U, Y_q is the tangent to the curve $y(t) = \phi_t q$ at $q = \phi_0 q$. This result can be proved almost exactly as in the classical case.

The new feature of the super setting is that corresponding to any odd vector field on $\mathcal{M}$ there is a local $(1,1)$ parameter super group of local transformations in a neighbourhood of any given point. Anticipating the full theory of super Lie groups presented in Chapter 9, and in particular Example 9.2.2, it is useful to observe that the set $\mathbb{R}_S^{1,1}$ has the structure of a group with

$$(t, \tau) \circ (s, \sigma) = (t + s + \tau\sigma, \tau + \sigma), \qquad (6.69)$$

since integral super curves lead to $(1, 1)$ parameter semigroups with this structure.

Definition 6.5.8   Let a be a positive real number and U be open in $\mathcal{M}$. A *local $(1, 1)$ parameter super group of local transformations* on $(\epsilon_{1,1})^{-1}(I_a) \times U$ is a mapping

$$\Phi : I_a \times U \to \mathcal{M} \qquad (6.70)$$

such that

(a) For each $(t, \tau) \in (\epsilon_{1,1})^{-1}(I_a)$ the mapping $\Phi_{t,\tau} : U \to \mathcal{M}, p \mapsto \Phi_{t,\tau}$ is a diffeomorphism of U onto its image.

(b)

$$\Phi_{t,\tau}(\Phi_{s,\sigma} p) = \Phi_{t+s+\sigma\tau, \tau+\sigma} p \qquad (6.71)$$

for t, s and p such that $\Phi_{s,\sigma} p \in U$ and s, t and $s + t$ are in I_a.

Theorem 6.5.9   *Suppose that Ξ is an odd vector field on $\mathcal{M}$ and p is a point in M. Then there exists a neighbourhood U of p, a positive real number a and a local $(1, 1)$ parameter super group of local transformations Φ on $(\epsilon_{1,1})^{-1}(I_a) \times U$ which induces Ξ in that at each q in U, Ξ_q is the odd tangent to the super curve $C(t, \tau) = \Phi_{t,\tau} q$ at $q = \Phi_{0,0} q$.*

Outline of proof   Suppose (V, ψ) is a coordinate chart on $\mathcal{M}$ containing p, with local coordinates $(X) = (x; \xi)$ such that $X(p) = 0$. Let $\Xi^i, i = 1, \ldots, m+n$ denote the components of Ξ. Consider the differential equation for $m + n$ functions $F^j, j = 1, \ldots, m + n$ on $\mathbb{R}_S^{1,1}$

$$\left(\frac{\partial^O}{\partial \tau} + \tau \frac{\partial^E}{\partial t} \right) F^j = \Xi^j(F(t, \tau)) \qquad (6.72)$$

Considering separately the component independent of τ and the component linear in τ leads to a system of ODE's which have a unique solution for some interval $|\epsilon(t)| < b$ corresponding to given initial condition $F(0, 0) = X$, provided that $X \in W$ where W is some sufficiently small neighbourhood of

$(0, 0)$ in $\mathbb{R}^{1,1}_S$. Let $F[X]$ denote this solution, and define $\Phi_{t,\tau} : \psi^{-1}(W) \to \mathcal{M}$ by

$$\Phi_{t,\tau}\, q = \psi^{-1}\left(F[X(q)](t;\tau)\right). \tag{6.73}$$

This map will be defined when $\epsilon(t)$ is in the interval I_c for some sufficiently small c. Since by inspection $F[X(q)](t + s + \tau\sigma, \tau + \sigma)$ satisfies (6.72) with initial condition $X(\Phi_{s,\sigma}\, q)$ it can be seen that Φ satisfies (6.71) whenever both sides are defined. Thus Φ satisfies the conditions of the theorem with $U = \psi^{-1}(W)$ and $a = \min(b, c)$. ∎

Chapter 7

Supermanifolds: The algebro-geometric approach

In this chapter the algebro-geometric approach to supermanifolds is described, in which it is a sheaf of functions over a manifold, rather than the manifold itself, which is extended. Section 7.1 gives the definition of a smooth algebro-geometric supermanifold, while in Section 7.2 local coordinates are discussed. The final section considers maps between supermanifolds.

This chapter is short because in the following chapter, essentially using a result first established by Batchelor [12], it is shown that smooth algebro-geometric supermanifolds correspond precisely to $(\mathbb{R}^{m,n}_S, \mathrm{DeWitt}, H^\infty)$ DeWitt supermanifolds (or, in the complex case, $(\mathbb{C}^{m,n}_S, \mathrm{DeWitt}, HC^\omega)$ supermanifolds) in such a way that there is a simple dictionary relating much of the differential geometry in the two approaches. So close is the correspondence that it is often unnecessary to state explicitly which approach is being used.

For the applications in this book the language of geometric supermanifolds is mostly more appropriate. Further material on the algebro-geometric approach may be found in [53, 39, 69]. As is the case with standard algebraic geometry, and also more general noncommutative geometry, ringed spaces and schemes play a major role; the generalised notion of points is obtained by considering ideals in the 'function' rings.

7.1 Algebro-geometric supermanifolds

The definition of supermanifold given below was given independently in broadly equivalent form by Berezin and Leĭtes [19] and Kostant [95]. First the notion of sheaf to be used will be defined. (What is called a sheaf here is sometimes called a complete presheaf in the literature; however in all cases used here, associated to any sheaf in the sense used here is a sheaf in the full sense; the association is canonical.)

Definition 7.1.1   Let X be a topological space.   A *presheaf A* of super commutative algebras over X is a collection of algebras $\{A(U)|U \text{ open in } X\}$ with the following properties.

(a) For each pair V, U of open sets in X such that $V \subset U$ there exists a *restriction map* $\rho_{U\,V}$ which is a super algebra homomorphism from $A(U)$ to $A(V)$.

(b) The restriction maps satisfy $\rho_{V\,W} \circ \rho_{U\,V} = \rho_{U\,W}$ whenever U, V and W are open in X and $W \subset V \subset U$.

The presheaf is said to be a *sheaf* if in addition the following properties hold for every open cover $\{U_\alpha | \alpha \in \Lambda\}$ of each open set U in X.

(a) If f, g are in $A(U)$, then $\rho_{U\,U_\alpha} f = \rho_{U\,U_\alpha} g$ for all $\alpha \in \Lambda$ implies that $f = g$.

(b) If $f_\alpha \in A(U_\alpha)$ is given for each $\alpha \in \Lambda$ with $\rho_{U_\alpha\,U_\alpha \cap U_\beta} f_\alpha = \rho_{U_\beta\,U_\alpha \cap U_\beta} f_\beta$ for all $\alpha, \beta \in \Lambda$, then there exists $f \in A(U)$ such that $f_\alpha = \rho_{U\,U_\alpha}$.

One example which is of relevance for this book is the sheaf C^∞ of smooth functions on a smooth manifold M. In this case, for any open set U in M, the algebra $C^\infty(U)$ is the algebra of smooth functions on this set, and the restriction maps are the usual restrictions of functions. A further example is the sheaf of smooth cross sections of a vector bundle over M. The notion of sheaf abstracts key features of the algebras of functions on a manifold, and allows a consistent generalisation to give the notion of algebro-geometric supermanifold.

Definition 7.1.2   A *smooth real algebro-geometric supermanifold* of dimension (m, n) is a pair (M, A) where M is a real m-dimensional manifold and A is a sheaf of super commutative algebras over M such that

(a) there exists an open cover $\{U_\alpha | \alpha \in \Lambda\}$ where for each α in Λ

$$A(U_\alpha) \cong C^\infty(U_\alpha) \otimes \Lambda(\mathbb{R}^n), \tag{7.1}$$

(b) if $\mathfrak{N}$ is the sheaf of nilpotents in A, then $(M, A/\mathfrak{N})$ is isomorphic to (M, C^∞).

A neighbourhood U in M on which $A(U)$ is isomorphic to $C^\infty(U) \otimes \Lambda(\mathbb{R}^n)$ is called a *splitting neighbourhood*.

A first example of an algebro-geometric supermanifold is the $(m,0)$-dimensional algebro-geometric supermanifold (M, C^∞) constructed from the sheaf of smooth functions on an m-dimensional manifold M. Less trivial examples of supermanifolds take the form $(M, \Gamma(\Lambda E))$ where E is a smooth vector bundle over a manifold M. (Here $\Gamma(\Lambda E)$ denotes the space of smooth cross-sections of the exterior bundle of E.) Such a supermanifold has dimension (m, n) where m is the dimension of M and n is the dimension of E, and is said to be split. The term split comes from the exact sequence of sheaves

$$0 \to \mathfrak{N}^2 \to A \to C^\infty \oplus E \to 0 \qquad (7.2)$$

which exists for any algebro-geometric supermanifold, with E here denoting the quotient sheaf $\mathfrak{N}/\mathfrak{N}^2$, which splits precisely when the supermanifold is derived from a vector bundle in this way. An important theorem due to Batchelor [11] shows that in fact any smooth algebro-geometric supermanifolds is (non-canonically) isomorphic to a split supermanifold. Under the identification that will be made of smooth algebro-geometric supermanifolds and H^∞ supermanifolds, Batchelor's theorem for algebro-geometric supermanifolds is implied by Theorem 8.2.1, which applies in the more general setting of concrete G^∞ supermanifolds.

7.2 Local coordinates on algebro-geometric supermanifolds

In this section (M, A) is an (m, n)-dimensional algebro-geometric supermanifold, and the notation of Definition 7.1.2 will be used. Harnessing the idea that elements of the super algebra $A(U)$ are to be regarded as functions on U in some generalised sense, it is possible to define coordinate neighbourhoods and even and odd coordinate systems on (M, A).

By part (b) of Definition 7.1.2, for each open subset U of M there is a homomorphism $\epsilon : A(U) \to C^\infty(U), f \mapsto f_{[\emptyset]}$, and these homomorphisms commute with restriction maps.

Suppose that U is a splitting neighbourhood of (M, A). Then there exist subalgebras $C(U)$ and $D(U)$ of $A(U)$ with $C(U) \cong C^\infty(U)$, $D(U) \cong \Lambda(\mathbb{R}^n)$ and

$$A(U_\alpha) = C(U_\alpha) \otimes D(U_\alpha). \qquad (7.3)$$

Since $\Lambda(\mathbb{R}^n)$ is generated by 1 and odd elements, the map $\epsilon|_{C(U_\alpha)} \to C^\infty$

must be an isomorphism. These properties make possible the following definition of coordinate neighbourhood and odd and even coordinate system.

Definition 7.2.1

(a) An open subset U of M is said to be a *coordinate neighbourhood* of (M, A) if it is both a coordinate neighbourhood of M and a splitting neighbourhood of (M, A).

(b) Suppose that U a coordinate neighbourhood of (M, A) with $A(U) = C(U) \otimes D(U)$ where $C(U) \cong C^\infty(U)$ and $D(U) \cong \Lambda(\mathbb{R}^n)$. Then an ordered set of generators $(\xi^1, \ldots; \xi^n)$ of $D(U)$ is said to be an *odd coordinate system* on U.

(c) A set $(x^1, \ldots, x^m)$ of even elements of $C(U)$ is said to be an *even coordinate system* on U if $(\epsilon x^1, \ldots, \epsilon x^m)$ is a coordinate system for M on U.

(d) A *coordinate system* of (M, A) on U consists of an even coordinate system and an odd coordinate system.

It is clear that any coordinate neighbourhood on (M, A) has at least one coordinate system. It can be shown that if $W \subset U$, and $(x^i; \xi^j), i = 1, \ldots, m; j = 1, \ldots, n$ is a coordinate system on U then $(\rho_{U\,W} x^i; \rho_{U\,W} \xi^j)$ is a coordinate system on W.

Suppose that U and V are coordinate neighbourhoods on (M, A) such that $U \cap V \neq \emptyset$, with coordinate systems $(x^i; \xi^j)$ and $(y^i; \eta^j)$, $i = 1, \ldots, m$, $j = 1, \ldots, n$ on U and V respectively. Then the restrictions of these coordinates to $U \cap V$ are also coordinates, and the restrictions of $(y^i; \eta^j)$ may be expressed in terms of those of $(x^i; \xi^j)$ in the following way:

$$y^j = \sum_{\underline{\mu} \in M_{N\,n0}} P^i_{\underline{\mu}}\, \xi^{\underline{\mu}}$$

$$\eta^j = \sum_{\underline{\mu} \in M_{N\,n1}} Q^j_{\underline{\mu}}\, \xi^{\underline{\mu}} \tag{7.4}$$

with each $P^i_{\underline{\mu}}$ and $Q^j_{\underline{\mu}}$ elements of $C(W)$. These explicit expressions can be used to show the connection between algebro-geometric and concrete supermanifolds [118].

Complex algebro-geometric supermanifolds can be defined in a closely analogous manner.

7.3 Maps between algebro-geometric supermanifolds

The notion of morphism of algebro-geometric supermanifolds is natural if one considers such a supermanifold to be a generalised function sheaf.

Definition 7.3.1 A smooth map h of a supermanifold (M, A) to a super-manifold (N, B) is a morphism of the sheaf A to the sheaf B such that the constituent algebra homomorphisms are super algebra homomorphisms.

In more detail this means that the map h consists of a C^∞ map $h_{[\emptyset]}$: $M \to N$ and, for each V open in N, a super algebra homomorphism h_V^* : $B(V) \to A(h^{-1}(V))$ which commutes with restriction maps in the sense that if $V' \subset V$ then

$$h_{V'}^* \circ \rho_{V\,V'} = \rho_{h^{-1}(V)\,h^{-1}(V')} \circ h_V^* . \tag{7.5}$$

In the trivial case where $A = C^\infty(M)$ and $B = C^\infty(N)$, any C^∞ map h from M to N induces a map of the supermanifolds (M, A) and (N, B), with h^* simply the standard pullback. Less trivially, if (M, A) has the form $(M, \Gamma(\Lambda E))$ and (N, B) has the form $(N, \Gamma(\Lambda F))$ where E and F are vector bundles over M and N respectively, then any vector bundle morphism $E \to F$ induces a smooth map of the supermanifolds (M, A) and (N, B). However, not all smooth maps between supermanifolds will be of this form. Helein has recently given an interesting account of maps in the algebro-geometric language [69]. An extended consideration of the supermanifold structure of the space of maps between supermanifolds is given by Batchelor in [13].

Chapter 8

The structure of supermanifolds

While there is no complete classification theorem for supermanifolds of all types, there are a number of cases where the classification problem can be reduced to a classification problem in classical differential geometry. In particular $(\mathbb{R}_S^{m,n}, \mathrm{DeWitt}, G^\infty)$ supermanifolds may be classified by smooth vector bundles. This chapter begins with a description of the construction of an $(\mathbb{R}_S^{m,n}, \mathrm{DeWitt}, G^\infty)$ supermanifold $\mathrm{S}(M, E)$ from an n-dimensional vector bundle E over an m-dimensional C^∞ manifold M. Analogous constructions for complex manifolds are also discussed. In section Section 8.2 it is shown (in a theorem originally due to Batchelor [11]) that any $(\mathbb{R}_S^{m,n}, \mathrm{DeWitt}, G^\infty)$ supermanifold $\mathcal{M}$ not only has a (non-canonical) $(\mathbb{R}_S^{m,n}, \mathrm{DeWitt}, H^\infty)$ structure, but also is superdiffeomorphic to a supermanifold of the form $\mathrm{S}(\mathcal{M}_{[\emptyset]}, E)$ where $\mathcal{M}_{[\emptyset]}$ is the body of $\mathcal{M}$ and the n-dimensional vector bundle E is uniquely determined by $\mathcal{M}$. This theorem is not true for complex analytic supermanifolds, as was originally demonstrated by Green [65]. An example to show this is given in Section 8.3.

Much of the structure of a supermanifold is contained in the rings of supersmooth functions considered in Section 6.1. The structure of the ring of H^∞ functions on a supermanifold suggests a correspondence with the abstract rings of the algebro-geometric approach. This correspondence is demonstrated in Section 8.4, where it is shown that the isomorphism classes of (m, n)-dimensional algebro-geometric supermanifolds correspond to the superdiffeomorphism classes of (m, n)-dimensional $(\mathbb{R}_S^{m,n}, \mathrm{DeWitt}, H^\infty)$ supermanifolds, or, in the complex case, $(\mathbb{C}_S^{m,n}, \mathrm{DeWitt}, HC^\omega)$ supermanifolds.

8.1 The construction of a split supermanifold from a vector bundle

Given an m-dimensional C^∞ manifold M together with an n-dimensional vector bundle E over M, one may use the transition functions of the manifold and the bundle to construct an (m, n)-dimensional $(\mathbb{R}_S^{m,n}, \mathrm{DeWitt}, H^\infty)$ supermanifold $S(M, E)$ with body M. This construction, which uses a patching technique, is described in the following theorem (where it is established that the object constructed is indeed a supermanifold).

Theorem 8.1.1   *Let M be a m-dimensional C^∞ manifold and let E be a C^∞ n-dimensional vector bundle over M. Suppose that $\{(U_\alpha, \phi_\alpha)|\alpha \in \Lambda\}$ is an atlas for M with coordinate neighbourhoods U_α which are also local trivialisation neighbourhoods of the bundle E. For each $\alpha, \beta \in \Lambda$ such that $U_\alpha \cap U_\beta \neq \emptyset$ let τ be the transition functions of M, that is,*

$$\tau_{\alpha\beta} = \phi_\alpha \circ \phi_\beta^{-1} : \phi_\beta(U_\alpha \cap U_\beta) \to \phi_\alpha(U_\alpha \cap U_\beta), \tag{8.1}$$

and

$$V_\alpha = (\epsilon_{m,n})^{(-1)}(\phi_\alpha(U_\alpha)) \subset \mathbb{R}_S^{m,n} \tag{8.2}$$

and also let

$$g_{\alpha\beta} : U_\alpha \cap U_\beta \to Gl(n, \mathbb{R}) \tag{8.3}$$

denote the transition functions of the bundle E.
   Now let

$$\mathcal{N} = \sqcup_{\alpha \in \Lambda} V_\alpha, \tag{8.4}$$

(were $\sqcup$ denotes disjoint union). Define

$$\check{\tau}_{\alpha\beta} \; : \; \epsilon_{m,n}{}^{(-1)}(\phi_\beta(U_\alpha \cap U_\beta)) \to \epsilon_{m,n}{}^{(-1)}(\phi_\alpha(U_\alpha \cap U_\beta))$$

by $\check{\tau}_{\alpha\beta}^i(x_\beta; \xi_\beta) = \widehat{\tau_{\alpha\beta}^i}(x_\beta) \qquad i = 1, \ldots, m$

$$\check{\tau}_{\alpha\beta}^{m+j}(x_\beta; \xi_\beta) = \sum_{l=1}^{n} \widehat{g_{\alpha\beta}{}^j{}_l \circ \phi_\beta^{-1}}(x_\beta)\xi_\beta^l \quad j = 1, \ldots, n. \tag{8.5}$$

Now let $\sim$ be the relation on $\mathcal{N} = \sqcup_{\alpha \in \Lambda} V_\alpha$ such that $(x; \xi) \sim (y; \eta)$ if and only if $\check{\tau}_{\alpha\beta}(x; \xi) = (y; \eta)$, where α and β are the unique elements of the index sets Λ such that $(y; \eta)$ is in V_α and $(x; \xi)$ is in V_β. Then

(a) The relation $\sim$ is an equivalence relation;

(b) The space $\mathcal{N}/\sim$ can be given the structure of an (m,n)-dimensional $(\mathbb{R}_S^{m,n}, \mathrm{DeWitt}, H^\infty)$ supermanifold.

(The supermanifold obtained from M and E in this way is denoted $\mathrm{S}(M,E)$.)

Proof

(a) Since $\tau_{\alpha\alpha}$ is the identity map on U_α and $g_{\alpha\alpha}$ maps U_α to the identity element of $Gl(n,\mathbb{R})$, the map $\check{\tau}_{\alpha\alpha}$ is the identity map on V_α, and thus $\sim$ is reflexive. Also, since the function $\tau_{\alpha\beta}$ is the inverse of the function $\tau_{\beta\alpha}$ and, for each x in $V_\alpha \cap V_\beta$, $g_{\alpha\beta}(x)$ is the matrix inverse of $g_{\beta\alpha[\emptyset]}(x)$, one finds that

$$\check{\tau}_{\beta\alpha} = (\check{\tau}_{\alpha\beta})^{-1}, \qquad (8.6)$$

and hence that the relation $\sim$ is symmetric. Finally, the transitivity of the relation follows from the fact that $\tau_{\alpha\beta} \circ \tau_{\beta\gamma} = \tau_{\alpha\gamma}$ on $\phi_\gamma(U_\alpha \cap U_\beta \cap U_\gamma)$ and the group product $g_{\alpha\beta}(x)g_{\beta\gamma}(x)$ is equal $g_{\alpha\gamma}(x)$ for all points x in $U_\alpha \cap U_\beta \cap U_\gamma$, so that

$$\check{\tau}_{\alpha\beta} \circ \check{\tau}_{\beta\gamma} = \check{\tau}_{\alpha\beta} \qquad (8.7)$$

on $(\epsilon_{m,n})^{(-1)}(\phi_\alpha(U_\alpha \cap U_\beta \cap U_\gamma))$.

(b) For each $p \in \mathcal{N}$ let $\bar{p}$ denote the equivalence class containing p and let $\bar{V}_\alpha$ denote the set of equivalence classes of points in V_α, and so on. Also define the map

$$\psi_\alpha : \bar{V}_\alpha \to \mathbb{R}_S^{m,n} \qquad \text{by}$$

$$\psi_\alpha^i(\bar{p}) = \widehat{\phi^i{}_\alpha}(x) \qquad i = 1,\ldots,m$$

$$\psi_\alpha^{m+j}(\xi) = \xi^j \qquad j = 1,\ldots,n$$

where $(x;\xi)$ is the unique element of the equivalence class $\bar{p}$ which lies in V_α. Then, since no two distinct points in V_α are equivalent under $\sim$, the map ψ_α is a bijective map of $\bar{V}_\alpha$ onto V_α, which is an open subset of $\mathbb{R}_S^{m,n}$ with the DeWitt topology. Also the transition function $\psi_\alpha \circ \psi_\beta : \psi_\beta(\bar{V}_\alpha \cap \bar{V}_\beta) \to \psi_\alpha(\bar{V}_\alpha \cap \bar{V}_\beta)$ is simply the map $\check{\tau}_{\alpha\beta}$ defined in equation (8.5). ∎

This construction can be adapted to give a complex supermanifold corresponding to a given complex vector bundle over an analytic manifold. In this case the resulting supermanifold is CH^ω.

A supermanifold which is superdiffeomorphic to a supermanifold constructed in this way is said to be split. (The terminology comes from the split exact sequence (7.2) which arises in the algebro-geometric formulation of such a supermanifold.) The atlas constructed above will not be maximal. Also, other atlases will exist on the supermanifold with transition functions of the form (8.5) ; any such atlas will be called a split atlas.

8.2 Batchelor's structure theorem for $(\mathbb{R}^{m,n}_S, \mathrm{DeWitt}, G^\infty)$ supermanifolds

In this section it will be shown that any $(\mathbb{R}^{m,n}_S, \mathrm{DeWitt}, G^\infty)$ supermanifold has a split structure. This result was first proved for H^∞ supermanifolds (in the algebro-geometric supermanifold formalism) by Batchelor [11]. The proof presented here is by no means the most elegant or mathematically sophisticated one possible; instead an explicit, hands-on technique for unravelling general G^∞ transition functions until they take take the split form (8.5) will be used. (This approach is similar to that used by Crane and Rabin when considering uniformisation of super Riemann surfaces [37].)

Theorem 8.2.1 *Any (m,n)-dimensional $(\mathbb{R}^{m,n}_S, \mathrm{DeWitt}, G^\infty)$ supermanifold $\mathcal{M}$ is split.*

Proof Let $\{(V_\alpha, \psi_\alpha)|\alpha \in \Lambda\}$ be the $(\mathbb{R}^{m,n}_S, \mathrm{DeWitt}, G^\infty)$ structure on the supermanifold $\mathcal{M}$. For each α, β such that $V_\alpha \cap V_\beta$ is not empty, let $T_{\alpha\beta}$ be the transition function

$$T_{\alpha\beta} = \psi_\alpha \circ \psi_\beta^{-1} : \Psi_\beta(V_\alpha \cap V_\beta) \to \Psi_\alpha(V_\alpha \cap V_\beta) \qquad (8.8)$$

Now each $T_{\alpha\beta}$ is invertible, and thus by Theorem 4.7.1 the superdeterminant of the matrix $\begin{pmatrix} \partial x^i_\alpha/\partial x^j_\beta & \partial x^i_\alpha/\partial \xi^j_\beta \\ \partial \xi^i_\alpha/\partial x^j_\beta & \partial \xi^i_\alpha/\partial \xi^j_\beta \end{pmatrix}$ is invertible. Thus, if for $k = 1, \ldots, n$

$$\xi^k_\alpha(x_\beta; \xi_\beta) = \gamma^k(x_\beta) + g_{\alpha\beta}{}^k{}_j(x_\beta)\xi^j_\beta + \text{higher order terms}, \qquad (8.9)$$

the function

$$g_{\alpha\beta} : \epsilon(\Psi_\beta(V_\alpha \cap V_\beta)) \to Gl(n, \mathbb{R})$$
$$\epsilon(x) \mapsto \left(g_{\alpha\beta}{}^k{}_j(\epsilon(x))\right) \qquad (8.10)$$

must be invertible at each point $(x_\beta; \xi_\beta)$ in the domain of $T_{\alpha\beta}$. Since the transition functions obey the compatibility condition

$$T_{\alpha\beta} \circ T_{\beta\gamma} = T_{\alpha\gamma} \tag{8.11}$$

on $\psi_\gamma(V_\alpha \cap V_\beta \cap V_\gamma)$, the functions $g_{\alpha\beta}$ are the transition functions of a uniquely determined n-dimensional real vector bundle E over the body $\mathcal{M}_{[\emptyset]}$ of $\mathcal{M}$. The aim of the proof will be to establish that on each coordinate neighbourhood V_α each even coordinate x_α^i can be replaced by an even coordinate $x_\alpha'^i$ such that

$$x'^i_\alpha(x'_\beta; \xi'_\beta) = \widehat{T^i_{[\emptyset]\alpha\beta}}(x'_\beta) \tag{8.12}$$

and each odd coordinate ξ_α^j can be replaced by an odd coordinate $\xi_\alpha'^k$ such that

$$\xi'^k_\alpha(x'_\beta; \xi'_\beta) = \widehat{g_{\alpha\beta}}(x'_\beta)^k{}_j \xi'^k_\beta \tag{8.13}$$

so that the supermanifold $\mathcal{M}$ is shown to have a split structure. These coordinate redefinitions are carried out order by order in ξ and order by order in the generators $\beta_{[r]}$ of the Grassmann algebra $\mathbb{R}_S$. Suppose that explicit expressions for the components of the transition functions $T_{\alpha\beta}$ are

$$T^i_{\alpha\beta}(x_\beta; \xi_\beta) = P^i_{\alpha\beta\,\underline{\mu}}(x_\beta)\xi_\beta^{\underline{\mu}} \qquad i = 1, \ldots, m$$
$$T^j_{\alpha\beta}(x_\beta; \xi_\beta) = Q^j_{\alpha\beta\,\underline{\mu}}(x_\beta)\xi_\beta^{\underline{\mu}} \qquad j = 1, \ldots, n$$

Then, since the transition functions $T_{\alpha\beta}$, $T_{\beta\gamma}$ and $T_{\alpha\gamma}$ obey the consistency conditions (8.11),

$$T^k_{\alpha\gamma}(x_\gamma; \xi_\gamma) = T^k_{\alpha\beta}(T_{\beta\gamma}(x_\gamma; \xi_\gamma)) \tag{8.14}$$

for all $(x_\gamma; \xi_\gamma)$ in $\psi_\gamma(V_\alpha \cap V_\beta)$ and for $k = 1, \ldots, m+n$. Each side of this equation is an element of $\mathbb{R}_S$, and thus having first equated coefficients in the expansion of each side in powers of ξ_γ, within each of these new equations coefficients of powers of the generators $\beta_{[r]}$ of $\mathbb{R}_S$ can be equated. Such coefficients will be labelled $_{\underline{\mu}[\underline{\rho}]}$ where the multi index $\underline{\mu}$ indicates that the coefficient of $\xi_\gamma^{\underline{\mu}}$ is being considered, while the $_{[\underline{\rho}]}$ indicates the coefficient of $\beta_{[\underline{\rho}]}$. It is in fact sufficient to consider points $(x_\gamma; \xi_\gamma)$ of the domain such that $s(x_\gamma) = 0$, since (by Definition 4.3.2) any functional relationship which holds at such points must extend to all points of the domain.

The first step is to consider the $\emptyset[\underline{\emptyset}]$ term in (8.11), which corresponds to the body $\mathcal{M}_{[\emptyset]}$ of $\mathcal{M}$, and gives

$$T_{[\emptyset]\alpha\gamma}{}^{i}(x_{\gamma};0) = T_{[\emptyset]\alpha\beta}{}^{i}(T_{[\emptyset]\beta\gamma}(x_{\gamma};0)) \tag{8.15}$$

which simply expresses the consistency of the transition functions of the body $\mathcal{M}_{[\emptyset]}$ of $\mathcal{M}$. The $\underline{\phi}[r]$ coefficient (with $[r]$ denoting the multi index containing the single element r) is the first to give useful information. In terms of the Q functions of (8.2) this coefficient gives

$$Q_{\alpha\gamma\underline{\emptyset}[r]}^{j}(x_{\gamma}) = Q_{\alpha\beta\underline{\emptyset}[r]}^{j}(x_{\beta}) + g_{\alpha\beta}(x_{\beta})^{j}{}_{\nu}Q_{\beta\gamma\underline{\emptyset}[r]}^{\nu}(x_{\beta}) \tag{8.16}$$

where x_{β} denotes the body of $T_{\beta\gamma}(x_{\gamma};\xi_{\gamma})$. This shows that $Q_{\alpha\gamma\underline{\emptyset}[r]}^{j}$ defines a 1-cycle in the Čech cohomology of $\mathcal{M}_{[\emptyset]}$ with coefficients in E. Now, essentially because of the existence of partitions of unity on $\mathcal{M}_{[\emptyset]}$, this sheaf is fine and thus has trivial q^{th} cohomology groups for $q \geq 1$. As a result there exists for each α in Λ and each natural number r a function

$$h_{\alpha[r]} : V_{[\emptyset]\alpha} \to \mathbb{R}^{n} \tag{8.17}$$

such that

$$Q_{\alpha\beta\emptyset[r]}^{j} = h_{\alpha[r]}^{j} - h_{\beta[r]}^{k}(g_{\alpha\beta})^{-1}{}_{k}{}^{j}(x_{\beta}) . \tag{8.18}$$

Thus the leading term $Q_{\alpha\beta\emptyset[r]}$ in $Q_{\alpha\beta\emptyset}$ can be set to zero by the G^{∞} redefinition of coordinate

$$\xi_{\alpha}^{j\,\prime} = \xi_{\alpha}^{j} + h_{\alpha[r]}^{j} . \tag{8.19}$$

Further coordinate redefinitions can then be made in a similar manner until all transition functions are reduced to the split form (8.5).   ∎

8.3   A non-split complex supermanifold

Batchelor's theorem, proved in the preceding section, showed that G^{∞} supermanifolds are always split. With complex analytic supermanifolds the situation is quite different; the proof of Batchelor's theorem is not valid, because the relevant cohomology groups do not vanish. Instead, the process can be turned on its head, and non-trivial cohomology elements used to construct explicitly non-split supermanifolds. The first examples of non-split supermanifolds were given by Green [65]. The more recent study of super Riemann surfaces (which are discussed in Chapter 14) provides many more examples of such supermanifolds.

The example of a non-split holomorphic supermanifold (first given by Green [65], using the algebro-geometric supermanifold formalism) will now be described. The supermanifold will be defined by specifying transition functions; the patching construction of Theorem 8.1.1 can then be used to construct the supermanifold.

Example 8.3.1   Let E denote the bundle $2T^*\mathbb{CP}^1$, that is, the double of the cotangent bundle of one-dimensional complex projective space $\mathbb{CP}^1$. Also, let e be a non-zero element of $H^1(T\mathbb{CP}^1 \otimes \Lambda^2 E)$. (Such an element must exist, since it is known that $H^1(T\mathbb{CP}^1 \otimes \Lambda^2 E) \simeq \mathbb{C}$.) Then, if $\{(U_\alpha, z_\alpha)|\alpha \in \Lambda\}$ is the holomorphic structure on $\mathbb{CP}^1$, for each α, β in Λ such that $U_\alpha \cap U_\beta$ is non-empty, there exists a nowhere zero function

$$e_{\alpha\beta} : U_\alpha \cap U_\beta \to \mathbb{C}^2$$

$$(8.20)$$

satisfying the cocycle condition

$$e_{\alpha\gamma} = e_{\alpha\beta} + e_{\beta\gamma} h_{\alpha\beta}^{-1} \qquad (8.21)$$

where

$$h_{\alpha\beta} = \partial z_\alpha / \partial z_\beta . \qquad (8.22)$$

The required supermanifold is constructed by taking coordinate patches $\{V_\alpha|\alpha \in \Lambda\}$ with

$$V_\alpha = (\epsilon^{1,2})^{-1}(z_\alpha(U_\alpha)) \qquad (8.23)$$

and transition functions

$$T_{\alpha\beta} : (\epsilon^{1,2})^{-1}(z_\alpha(U_\alpha \cap U_\beta)) \to (\epsilon^{1,2})^{-1}(z_\beta(U_\alpha \cap U_\beta))$$

$$\text{with } T_{\alpha\beta}^1(z, \zeta) = \widehat{\phi_{\alpha\beta}}(z) + e_{\alpha\beta}\epsilon_{ij}\zeta^i\zeta^j$$

$$T_{\alpha\beta}^{1+j}(z, \zeta) = (\widehat{h_{\alpha\beta}}(z))^{-1}\zeta^j, \quad j = 1, 2 \qquad (8.24)$$

where $\phi_{\alpha\beta} : z_\alpha(U_\alpha \cap U_\beta) \to z_\beta(U_\alpha \cap U_\beta)$ are the transition functions of $\mathbb{CP}^1$. It follows from the cocycle condition (8.21) that the transition functions $T_{\alpha\beta}$ obey the consistency condition (8.7), and hence the disjoint union of the V_α may be patched together to define a supermanifold, which in this case is non-split.

Rothstein has shown that although not all complex supermanifolds are split, in the H^∞ case they can be regarded as deformations of split supermanifolds[132].

8.4   Comparison of the algebro-geometric and concrete approach

In this section the correspondence between algebro-geometric and concrete supermanifolds is demonstrated. First, the existence of a unique algebro-geometric supermanifold corresponding to a given H^∞ manifold will be shown.

Proposition 8.4.1   *Let $\mathcal{M}$ be an H^∞ supermanifold of dimension (m, n), and let A be the sheaf of super algebras on $\mathcal{M}_{[\emptyset]}$ with*

$$A(V) = H^\infty(\epsilon^{-1}(V)). \tag{8.25}$$

Then $(\mathcal{M}_{[\emptyset]}, A)$ is an algebro-geometric supermanifold of dimension (m, n).

Outline of proof A is a sheaf for the usual reasons for sheaves of functions. Let $\{(V_\alpha, \psi_\alpha) | \alpha \in \Lambda\}$ be an atlas of H^∞ charts on $\mathcal{M}$ and let $U_\alpha = \epsilon(V_\alpha)$ for each α in Λ. Then it follows from Definition 4.4.3 that

$$A(U_\alpha) = H^\infty(V_\alpha) \cong C^\infty(U_\alpha) \otimes \Lambda(\mathbb{R}^n). \tag{8.26}$$

It is clear from the same definition that for each open U in $\mathcal{M}_{[\emptyset]}$,

$$A(U)/\mathfrak{N}(U) \cong C^\infty(U). \tag{8.27}$$

∎

    There is an analogous result for complex algebro-geometric supermanifolds [4]. Conversely, given any algebro-geometric supermanifold (X, A), it is possible to construct a supermanifold $H(X, A)$ of matching dimension such that the body of $H(X, A)$ is X, and so that the algebro-geometric supermanifold corresponding to the sheaf of H^∞ functions on $H(X, A)$ is isomorphic to (X, A). Moreover the correspondence now constructed between algebro-geometric supermanifolds and $(\mathbb{R}_S^{m,n}, \text{DeWitt}, H^\infty)$ supermanifolds is bijective. One proof of this result is given in [12]. Alternatively the result may be established by constructing the concrete supermanifold corresponding to a given algebro-geometric supermanifold using a patching technique, and the explicit form (7.4) of the coordinate transition functions for an algebro-geometric supermanifold [118]. The process is very similar to that used in Theorem 8.1.1 to construct the supermanifold $S(M, E)$, and so will not be described in detail.

Chapter 9

Super Lie groups

One of the principal motivations for studying supermanifolds is the desire to find the appropriate global object or 'super Lie group' corresponding to a super Lie algebra. In supersymmetric physics elements of a super Lie algebra are seen as infinitesimal generators of transformation, with anticommuting parameters attached to the odd elements and commuting parameters to the even ones. Comparing this to the case of conventional Lie groups, where the parameters of the infinitesimal generators can be used as coordinates on a neighbourhood of the identity, this suggests that a super Lie group will have both commuting and anticommuting local coordinates, and thus that a super Lie group should be a supermanifold. This leads naturally, by analogy with the definition of a conventional Lie group, to the definition of a super Lie group as a group which is also a supermanifold, with supersmooth group operations. Such an approach is explored in detail in this chapter, and does indeed provide the appropriate mathematical framework for the groups used in supersymmetric theories. Moreover the relationship between a super Lie group and its super Lie algebra is shown to take the expected form, with one or two extra possibilities. Because supermanifolds are topological spaces, the formulation of super Lie groups given here naturally incorporates global topological properties.

Super Lie groups derived from graded Lie algebras were first considered by Berezin and Kac, in a mathematical context, some years before the physical theories mentioned above; in their paper 'Lie Groups with commuting and anticommuting parameters' [18] they give a definition of a formal super Lie group. Subsequently, Kostant has given a full account of the notion of 'graded Lie group' in his extensive work on algebro-geometric supermanifolds [95]. In this approach it is once again the function algebra which is extended, the extension from group to super group is made using

Lie Hopf algebras. Neither the formal groups of Berezin and Kac nor the graded Lie groups of Kostant are actually abstract groups, in contrast to the super Lie groups defined in this chapter. These super Lie groups bear the same relationship to the formal groups of Berezin and Kac as do conventional Lie groups to the formal Lie groups first introduced by Bochner [25]. Fourier analysis on super Lie groups has been developed and applied by Zirnbauer[163] and by Hüffmann[80].

Section 9.1 gives the basic definition of super Lie group, and introduces the corresponding super Lie module. Examples of super Lie groups are given in Section 9.2, while in Sections 9.3 and 9.4 the correspondence between super Lie groups, super Lie modules and super Lie algebras is established. Finally, in Section 9.5 it is shown that the graded Lie groups of Kostant can be related to the super Lie group formulation given here.

9.1   The definition of a super Lie group

The basic definition of a super Lie group [120] closely parallels that of a conventional Lie group.

Definition 9.1.1   A *super Lie group* G is a group G which also has the structure of a $(G^\omega, \mathrm{DeWitt}, \mathbb{R}_S^{m,n})$ supermanifold, with the group operations

$$
\begin{aligned}
G \times G &\to G, & (g_1, g_2) &\mapsto g_1 g_2, \\
\text{and} \quad G &\to G, & g &\mapsto g^{-1}
\end{aligned}
\tag{9.1}
$$

being G^ω.

This definition can be generalised to include groups which are supermanifolds modelled on other super algebras.

It is a well known result of classical Lie theory that the left invariant vector fields on a Lie group (sometimes referred to as infinitesimal right translations) form a Lie algebra under commutation, and that this algebra characterises the local properties of the group. The analogous property of a super Lie group is that the left invariant vector fields form a super Lie module (Definition 2.4.3); in almost all cases this super Lie module has the product structure $\mathbb{A} \otimes \mathfrak{g}$ of Example 2.4.5, with $\mathbb{A}$ the super algebra on which the super Lie group is modelled (as a supermanifold) and $\mathfrak{g}$ a super Lie algebra. Although the definition is exactly as in the classical case, for completeness the notion of left invariant vector field will now be defined.

Definition 9.1.2   Let G be a super Lie group and let $g \in G$.

(a) Define the *left action* δ_g of g on G to be the mapping

$$\delta_g : G \to G \qquad g' \mapsto gg'. \tag{9.2}$$

(b) Define δ_{g*} to be the induced mapping on vector fields on G, so that

$$\delta_{g*}(X)f = X(f \circ \delta_g) \quad \text{for all} \quad f \in G^\infty(G), X \in \mathcal{D}(G). \tag{9.3}$$

(c) A vector field X in $\mathcal{D}(G)$ is said to be *left invariant* if

$$\delta_{g*}(X) = X \quad \text{for all} \quad g \in G. \tag{9.4}$$

(d) The set of left invariant vector fields on G is denoted $\mathcal{L}(G)$.

Equipped with this definition, the proof of the main result of this section proceeds very much as in the classical case.

Theorem 9.1.3 *Suppose that G is an (m, n)-dimensional super Lie group. Then $\mathcal{L}(G)$ is an (m, n)-dimensional super Lie module over $\mathbb{R}_S$ under the bracket operation*

$$[,] : \mathcal{L}(G) \times \mathcal{L}(G) \to \mathcal{L}(G)$$
$$[X, Y] = XY - (-1)^{|X||Y|}YX.$$

Proof It is proved in Theorem 6.4.3 that $\mathcal{D}(G)$, the space of vector fields on G, forms a super $\mathbb{R}_S$-module under the bracket operation defined above. It can be shown, using an analogue of the classical proof, that $\mathcal{L}(G)$ is a Lie sub module, in particular that it is closed under the bracket operation.

   Let $T_e(G)$ be the tangent space at the identity of G. Then $T_e(G)$ is an (m, n)-dimensional super vector space. Also the mapping

$$\mathcal{L}(G) \to T_e(G), \qquad X \mapsto X_e \tag{9.5}$$

can be shown to be a super vector space isomorphism, using an analogue of the classical proof. ∎

In almost all cases the super Lie module of a super Lie group has the product structure $\mathbb{R}_S \otimes \mathfrak{g}$ where $\mathfrak{g}$ is a super Lie algebra, and so one can refer to the super Lie algebra of a super Lie group.

9.2　Examples of super Lie groups

In this section, several examples of super Lie groups and their super Lie modules will be described. First, for completeness, the Abelian translation super Lie groups are described. While these may seem to be the natural analogue of the standard translation groups, it is the 'supertranslation' groups defined in Example 9.2.2 which are much more significant and characteristic of supersymmetry. A supertranslation group can also be combined with the Lorentz group as a semi direct product, giving the super Poincaré group. There are analogues of the classical Lie groups, some of which are described below. A more complete account of these groups and their super Lie algebras is given in [136] and [34]. Following this, a generic construction of a super Lie group corresponding to an arbitrary super Lie algebra is given; the resulting super Lie group is the semidirect product of a standard Lie group and a nilpotent Lie group. This construction is used in Section 9.5 to relate the concrete super Lie groups to those of the algebro-geometric approach. Finally, as a curiosity, two super Lie groups are presented in Example 9.2.3 whose super Lie module does not factorise into the tensor product of a super Lie algebra and a super algebra.

The first example of a super Lie group simply shows that superspace $\mathbb{R}_S^{m,n}$ itself has the structure of an additive group:

Example 9.2.1 The space $\mathbb{R}_S^{m,n}$ is an (m,n)-dimensional abelian super Lie group with group action defined by

$$(x;\xi) + (y;\eta) = (x+y;\xi+\eta). \qquad (9.6)$$

An example of a non-abelian super Lie group, which plays an important rôle in supersymmetric theories is the group of super translations, which will now be described.

Example 9.2.2 Suppose that m and n are integers such that $SO(1,m-1)$ has Majorana spinors of dimension n, and that $\gamma^m{}_\alpha{}^\beta, \alpha,\beta = 1,\ldots,n$, $i = 1,\ldots m$ are Dirac matrices satisfying

$$\sum_{\beta=1}^{n} \left(\gamma^i{}_\alpha{}^\beta \gamma^j{}_\beta{}^\sigma + \gamma^j{}_\alpha{}^\beta \gamma^i{}_\beta{}^\sigma \right) = \delta^\sigma_\alpha \eta^{ij}, \quad \alpha,\sigma = 1,\ldots,n,\ i,j = 1,\ldots,m$$

$$(9.7)$$

where η is the Minkowski metric on $\mathbb{R}^m$ preserved by $SO(m-1,1)$. Then

$\mathbb{R}_S^{m,n}$ is given the structure of a super Lie group by defining

$$(x^1, \dots, x^m; \theta^1, \dots, \theta^n) \circ (y^1, \dots, y^m; \phi^1, \dots, \phi^n)$$
$$= (z^1, \dots, z^m; \psi^1, \dots, \psi^n) \tag{9.8}$$

with

$$z^i = x^i + y^i + \sum_{\alpha,\beta=1}^{n} \theta^\alpha \phi^\beta (C\gamma^i)_{\alpha\beta}, \quad i = 1, \dots m \quad \text{and}$$
$$\psi^\alpha = \theta^\alpha + \phi^\alpha, \qquad \alpha = 1, \dots n. \tag{9.9}$$

This group will be denoted $T^{m,n}$.

A simple example of this group is $T^{1,1}$ which has group law

$$(s; \sigma) \circ (t; \tau) = (s + t + \sigma\tau; \sigma + \tau). \tag{9.10}$$

As with the simple translation group, there is a natural coordinate system on $T^{m,n}$ and it is almost immediate that this gives the supertranslation group the structure of a super Lie group; using the construction described below, it can be shown that the super Lie algebra of this group is the (m, n)-dimensional super algebra with generators $P_i, i = 1, \dots, m$ and $Q_\alpha, \alpha = 1, \dots, n$ and non-zero brackets

$$[Q_\alpha, Q_\beta] = \sum_{i=1}^{m} (C\gamma^i)_{\alpha\beta} P_i. \tag{9.11}$$

This is the basic $N = 1$ supersymmetry algebra in m dimensions. In supersymmetric physical theories, which are considered in Chapter 13, it is useful also to include Lorentz rotations, by talking the semi direct product with the supersymmetry algebra using the vector representation for the even part and the spinor representation for the odd part. This leads to the *super Poincaré algebra* which has a further $\frac{1}{2}m(m-1)$ even generators $J_{ij}, i, j = 1, \dots, m$ with $J_{ij} = -J_{ji}$ and brackets

$$[J_{ij}, Q_\alpha] = -\tfrac{1}{2}\Sigma_{ij}{}_\alpha{}^\beta Q_\beta$$
$$[J_{ij}, P_k] = \eta_{ik} P_i - \eta_{jk} P_i$$
$$[J_{ij}, J_{kl}] = \eta_{jk} J_{il} - \eta_{ik} J_{jl} - \eta_{jl} J_{ik} + \eta_{il} J_{jk} \tag{9.12}$$

where $\Sigma_{ij}{}_\alpha{}^\beta = \frac{1}{2}\left(\gamma_{i\alpha}{}^\gamma \gamma_{j\gamma}{}^\beta - \gamma_{j\alpha}{}^\gamma \gamma_{i\gamma}{}^\beta\right)$. The corresponding super Lie group is the semi direct product of the supertranslation group $T^{m,n}$ and the spin double cover of $SO(m-1, 1)$.

Matrix super Lie groups and their super Lie algebras have been considered by a number of authors, for example Rittenberg [116], Scheunert [136] and Cornwell [34]. The basic example is the group $GL(m, n; \mathbb{R}_S)$ of invertible $(m, n) \times (m, n)$ super matrices, which is an $(m^2 + n^2, 2mn)$-dimensional super Lie group. (The proof that this is a super Lie group is very much as in the classical case.) It has many super Lie subgroups, defined by various restrictions. One example is the special linear group $SL(m, n; \mathbb{R}_S)$, which consists of matrices in $GL(m, n; \mathbb{R}_S)$ whose superdeterminant is equal to 1. Another example is the *orthosymplectic* group $OSP(m, l; \mathbb{R}_S$ which consists of $(m, 2l) \times (m, 2l)$ super matrices M which preserve the flat 'super metric'

$$\Upsilon = \begin{pmatrix} 1_m & 0 \\ 0 & C_l \end{pmatrix} \tag{9.13}$$

(where 1_m denotes the $m \times m$ identity matrix and C_l is the $2l \times 2l$ standard symplectic matrix with l copies of $\begin{pmatrix} 0 & 1 \\ -1 & 0 \end{pmatrix}$ down the leading diagonal) in the sense that

$$M^{ST} \Upsilon M = \Upsilon. \tag{9.14}$$

Two non-Abelian super Lie groups are now presented which have the same super Lie module $\mathfrak{u}$, but are topologically distinct in both the even and the odd sector. In this example a finite-dimensional Grassmann algebra $\mathbb{R}_{S[4]}$ is used, and also for the second example the fine topology. An additional feature is that the super Lie module $\mathfrak{u}$ does not factorise as $\mathfrak{g} \otimes \mathbb{R}_{S[4]}$ for any super Lie algebra $\mathfrak{g}$.

Example 9.2.3

(a) Let $H = \mathbb{R}_{S[4]}^{1,1}$ with group operation defined by

$$(a; \alpha) \circ (c; \gamma) = (a + c + \tfrac{1}{2}\beta_{[1]}\beta_{[2]}\alpha\gamma; \alpha + \gamma). \tag{9.15}$$

This super Lie group is $(1, 1)$-dimensional and has super Lie module which can be expressed in terms of a super basis $\{Y_1, Y_2\}$ with brackets

$$[Y_1, Y_1] = [Y_1, Y_2] = 0, \quad \text{and} \quad [Y_2, Y_2] = \beta_{[1]}\beta_{[2]}Y_1. \tag{9.16}$$

(b) Let $G = H/D$ where D is the discrete central subgroup of H consisting of elements of the form

$$\left(\sum_{\underline{\mu} \in M_{4,0}} m^{\underline{\mu}} \beta_{[\underline{\mu}]}, \right.$$

$$n^1 \beta_{[1]} + n^2 \beta_{[2]} + n^{123} \beta_{[1]} \beta_{[2]} \beta_{[3]} + n^{134} \beta_{[1]} \beta_{[3]} \beta_{[4]}$$

$$\left. + n^{124} \beta_{[1]} \beta_{[2]} \beta_{[4]} + n^{234} \beta_{[2]} \beta_{[3]} \beta_{[4]} \right) \qquad (9.17)$$

where each $m^{\underline{\mu}}$ and $n^{\underline{\nu}}$ is an integer. G is then a super Lie group (with supermanifold topology the fine topology) with the same super Lie module as H, but homeomorphic to $(S^1)^6 \times \mathbb{R}^{10}$ rather than to $\mathbb{R}^{16}$.

There are super Lie groups which are $(G^\infty, \mathbb{R}_S^{m,n}, \text{DeWitt})$ supermanifolds and have this same super Lie module structure (but over $\mathbb{R}_S$). However they will be the analogue of (a) in the preceding example rather than (b) which requires the fine topology.

9.3   The construction of a super Lie group with given super Lie $\mathbb{R}_{S[L]}$-module

In Section 9.1 it was shown that each super Lie group has an associated super Lie module. In this section the first step towards a converse result is established, that is, it is shown that corresponding to any super Lie module $\mathfrak{u}$ over $\mathbb{R}_{S[L]}$, with L greater than the odd dimension of the module, there is a super Lie group (not necessarily unique) whose super Lie module is $\mathfrak{u}$. The proof also shows that the fine structure manifold of the super Lie group is a Lie group whose Lie algebra is the fine structure Lie algebra of $\mathfrak{u}$.

The technique employed to establish the main theorem of this section is to relate structures on $\mathbb{R}_{S[L]}^{m,n}$ to those on $\mathbb{R}^{2^{L-1}(m+n)}$ via the identification

$$\iota : \mathbb{R}_{S[L]}^{m,n} \to \mathbb{R}^{2^{L-1}(m+n)} \qquad (9.18)$$

$$\left(\sum_{\underline{\mu}_1 \in M_{L,0}} x^{1\underline{\mu}} \beta_{[\underline{\mu}_1]}, \dots, \sum_{\underline{\mu}_m \in M_{L,0}} x^{m\underline{\mu}} \beta_{[\underline{\mu}_m]}; \right.$$

$$\left. \sum_{\underline{\nu}_{m+1} \in M_{L,1}} x^{(m+1)\underline{\nu}} \beta_{[\underline{\nu}_{m+1}]}, \dots, \sum_{\underline{\nu}_{m+n} \in M_{L,1}} x^{(m+n)\underline{\nu}} \beta_{[\underline{\nu}_{m+n}]} \right)$$

$$\mapsto \left(x^{1\underline{\mu}}, \dots, x^{m\underline{\mu}}, \; x^{(m+1)\underline{\nu}}, \dots, x^{(m+n)\underline{\nu}} \,|\, \underline{\mu} \in M_{L,0}, \underline{\nu} \in M_{L,1} \right).$$

$$(9.19)$$

The following lemma establishes a criterion by which it may be determined whether or not a given analytic function on $\mathbb{R}^{2^{L-1}(m+n)}$ may be identified with a superanalytic function on $\mathbb{R}^{m,n}_{S[L]}$. The conditions of this lemma were used by Boyer and Gitler to characterise supersmooth functions [26].

Lemma 9.3.1 *Suppose that U is open in $\mathbb{R}^{2^{L-1}(m+n)}$ and that $f : U \to \mathbb{R}_{S[L]}$ is analytic. Then $f \circ \iota$ is superanalytic on $\iota^{-1}(U)$ if an only if for every finite sequence $p_1, p_2, \ldots, p_k$ of positive integers with each p_r less than or equal to $m + n$ there exist elements $f_{p_1 \ldots p_k}$ of $\mathbb{R}_{S[L]}$ such that*

$$\partial_{p_k \underline{\mu}_k} \cdots \partial_{p_1 \underline{\mu}_1} f(0) = \beta_{[\underline{\mu}_1]} \cdots \beta_{[\underline{\mu}_k]} f_{p_1 \ldots p_k} \tag{9.20}$$

for each $\underline{\mu}_r \in M_{L,|p_k|}, r = 1, \ldots, k$.

This lemma, which provides the analogue of the Cauchy Riemann equations, may be proved by considering the series expansion about zero. The following corollary can be useful.

Corollary 9.3.2

$$f_{p_1 \ldots p_k} = \partial^S_{p_k} \cdots \partial^S_{p_1}(f \circ \iota)(0). \tag{9.21}$$

The next lemma shows how, given a Lie group whose Lie algebra is equal to the even part of a super Lie module, the local analytic structure of this Lie group may be used to construct a local superanalytic structure.

Lemma 9.3.3 *Let $\mathfrak{u}_{[L]}$ be an (m,n)-dimensional super Lie module over $\mathbb{R}_{S[L]}$, and $\{X_i | i = 1, \ldots, m + n\}$ be a super basis of $\mathfrak{u}_{[L]}$. Also let $\mathfrak{h}$ denote the even part of $\mathfrak{u}$ regarded as a $2^{L-1}(m + n)$-dimensional Lie algebra, so that $\mathfrak{h}$ has a basis*

$$\left\{ X_{i\underline{\mu}} | X_{i\underline{\mu}} = \beta_{[\underline{\mu}]} X_i, i = 1, \ldots, m + n, \underline{\mu} \in M_{L,|i|} \right\}. \tag{9.22}$$

Let H be a Lie group whose Lie algebra is $\mathfrak{h}$, and let $\phi_e : V \to \mathbb{R}^{2^{L-1}(m+n)}$ be canonical coordinates (with respect to the basis $\{X_{i\underline{\mu}}\}$) on some neighbourhood V of the identity e of H. Let U be a neighbourhood of e such that $UU \subset V$ and let the analytic function k be defined by

$$k : \phi_e(U) \times \phi_e(U) \to \phi_e(V), \quad (\phi_e(g), \phi_e(h)) \mapsto \phi_e(gh). \tag{9.23}$$

(The $(i\underline{\mu})$ component of this map will be denoted $k^{i\underline{\mu}}$.) Also let the function $\psi_e : V \to \mathbb{R}^{m,n}_{S[L]}$ be defined by

$$\psi_e = \iota^{-1} \circ \phi_e. \tag{9.24}$$

Suppose that $K : \psi_e(U) \times \psi_e(U) \to \psi_e(V)$ is defined by

$$K^i(x,y) = \sum_{\underline{\mu} \in M_{L|i|}} k^{i\underline{\mu}} \left(\iota(x), \iota(y) \right) \beta_{[\underline{\mu}]} . \tag{9.25}$$

Then

(a) $K\left(\psi_e(g), \psi_e(h)\right) = \psi_e(gh)$,
(b) K is super analytic, and
(c) if for $i, j = 1, \ldots, m+n$ functions $\chi_j^i : \psi_e(U) \to \mathbb{R}_{S[L]}$ are defined by

$$\chi_j^i(x) = \partial^S_{jy} K^i \big|_{y=0} \tag{9.26}$$

then

$$\partial^S_i \chi_j^k(0) = \frac{1}{2} C^k_{ij\,[L]} , \tag{9.27}$$

where C^k_{ij} are structure constants for the basis $\{X_i | i = 1, \ldots, m+n\}$ of $\mathfrak{u}_{[L]}$.

(Properties (a) and (b) mean that group operations expressed in terms of the chart (ψ_e, V) are superanalytic.)
Proof (a)

$$
\begin{aligned}
K^i\left(\psi_e(g), \psi_e(h)\right) &= \sum_{\underline{\mu} \in M_{L|i|}} k^{i\underline{\mu}} \left(\iota \circ \psi_e(g), \iota \circ \psi_e(h) \right) \beta_{[\underline{\mu}]} \\
&= \sum_{\underline{\mu} \in M_{L|i|}} \phi_e^{i\underline{\mu}}(gh) \beta_{[\underline{\mu}]} \\
&= \psi_e^i(gh) .
\end{aligned} \tag{9.28}
$$

(b) For $i, j = 1, \ldots, m+n$, $\underline{\mu} \in M_{L|i|}$ and $\underline{\nu} \in M_{L|j|}$, let

$$\chi_{j\underline{\nu}}^{i\underline{\mu}} = \frac{\partial k^{i\underline{\mu}}}{\partial y^{j\underline{\nu}}}(x,y) \bigg|_{y=0} , \tag{9.29}$$

where $x, y \in \phi_e(V)$. Then, using Lie's first theorem, it will be shown by induction over r that, for any finite sequence $(p_1, \ldots, p_r)$ of positive integers between 1 and $m+n$ inclusive, and multi-indices $\underline{\sigma}_s \in M_{L\,|p_s|}, s = 1, \ldots, r$,

$$
\frac{\partial k^{i\underline{\mu}}(x,y)}{\partial y^{p_r \underline{\sigma}_r} \ldots \partial y^{p_1 \underline{\sigma}_1}}
$$

$$
= \sum_{a=1}^{m+n} \sum_{\underline{\alpha} \in M_L} \sum_{s_1=1}^{m+n} \cdots \sum_{s_r=1}^{m+n} \sum_{\underline{\tau}_1 \in M_{L|s_1|}, \underline{\gamma}_1 \in M_L} \cdots \sum_{\underline{\tau}_r \in M_{L|s_r|}, \underline{\gamma}_r \in M_L}
$$

$$
T(r)_{s_1 \ldots s_r}^{a \underline{\gamma}_1 \cdots \underline{\gamma}_r} \, F_{\underline{\tau}_1 \ldots \underline{\tau}_r \underline{\gamma}_1 \cdots \underline{\gamma}_r}^{\underline{\alpha}} \, \chi_{a\underline{\alpha}}^{i\underline{\mu}}(\kappa(x,y)) \, \tilde{\chi}_{p_1 \underline{\sigma}_1}^{s_1 \underline{\tau}_1}(x,y) \ldots \tilde{\chi}_{p_r \underline{\sigma}_r}^{s_r \underline{\tau}_r}(x,y)
$$

$$(9.30)$$

where $T(r)_{s_1 \ldots s_r}^{a\underline{\gamma}_1 \cdots \underline{\gamma}_r}$ is independent of x and y, $F_{\underline{\tau}_1 \ldots \underline{\tau}_r \underline{\gamma}_1 \cdots \underline{\gamma}_r}^{\underline{\alpha}}$ is defined by

$$
\beta_{[\underline{\nu}_1]} \ldots \beta_{[\underline{\nu}_k]} = \sum_{\underline{\alpha} \in M_L} F_{\underline{\nu}_1 \ldots \underline{\nu}_k}^{\underline{\alpha}} \beta_{[\underline{\alpha}]}, \tag{9.31}
$$

and $\tilde{\chi}$ is the inverse of the matrix χ.

To establish this result, note that Lie's first theorem states that

$$
\frac{\partial k^{i\underline{\mu}}}{\partial y^{p_s \underline{\sigma}_s}}(x,y) = \sum_{a=1}^{m+n} \sum_{\underline{\alpha} \in M_L} \chi_{a\underline{\alpha}}^{i\underline{\mu}}(k(x,y)) \tilde{\chi}_{p_s \underline{\sigma}_s}^{a\underline{\alpha}}(y). \tag{9.32}
$$

Thus (9.30) holds when $r = 1$ with $T(1)_{s_1}^{a\underline{\gamma}_1} = \delta_{s_1}^{a} \delta_{\emptyset}^{\underline{\gamma}_1}$.

Now, since ϕ_e is a canonical coordinate with respect to the basis $\{X_{i\underline{\mu}}\}$ of $\mathfrak{h}$,

$$
X_{i\underline{\mu}} = \sum_{j=1}^{m+n} \sum_{\underline{\nu} \in M_L} \chi_{i\underline{\mu}}^{j\underline{\nu}} \partial_{j\underline{\nu}}, \tag{9.33}
$$

and thus, if $\left[X_{i\underline{\mu}}, X_{j\underline{\nu}} \right] = B_{i\underline{\mu}\,j\underline{\nu}}^{k\underline{\sigma}} X_{k\underline{\sigma}}$,

$$
B_{i\underline{\mu}\,j\underline{\nu}}^{k\underline{\sigma}} = \left(\chi_{i\underline{\mu}}^{h\underline{\nu}} \partial_{h\underline{\nu}} \chi_{j\underline{\nu}}^{l\underline{\lambda}} - \chi_{j\underline{\nu}}^{h\underline{\nu}} \partial_{h\underline{\nu}} \chi_{i\underline{\mu}}^{l\underline{\lambda}} \right) \tilde{\chi}_{l\underline{\lambda}}^{k\underline{\sigma}}
$$

$$
= \left(-\chi_{i\underline{\mu}}^{h\underline{\nu}} \chi_{j\underline{\nu}}^{l\underline{\lambda}} + \chi_{j\underline{\nu}}^{h\underline{\nu}} \chi_{i\underline{\mu}}^{l\underline{\lambda}} \right) \partial_{h\underline{\nu}} \tilde{\chi}_{l\underline{\lambda}}^{k\underline{\sigma}}. \tag{9.34}
$$

But $\partial_{h\underline{\nu}} \tilde{\chi}_{l\underline{\lambda}}^{k\underline{\sigma}} = -\partial_{l\underline{\lambda}} \tilde{\chi}_{h\underline{\nu}}^{k\underline{\sigma}}$, so that

$$
B_{i\underline{\mu}\,j\underline{\nu}}^{k\underline{\sigma}} = -2\chi_{i\underline{\mu}}^{h\underline{\nu}} \chi_{j\underline{\nu}}^{l\underline{\lambda}} \partial_{h\underline{\nu}} \tilde{\chi}_{l\underline{\lambda}}^{k\underline{\sigma}}. \tag{9.35}
$$

This gives

$$\partial_{h\underline{\nu}}\chi^{k\underline{\sigma}}_{l\underline{\lambda}}(x) = \frac{1}{2}\tilde{\chi}^{i\underline{\mu}}_{h\underline{\nu}}(x)\chi^{k\underline{\sigma}}_{r\underline{\rho}}(x)B^{r\underline{\rho}}_{i\underline{\mu}\, l\underline{\lambda}} \tag{9.36}$$

and

$$\partial_{h\underline{\nu}}\chi^{k\underline{\sigma}}_{l\underline{\lambda}}(k(x,y)) = \frac{1}{2}\tilde{\chi}^{i\underline{\mu}}_{h\underline{\nu}}(y)\chi^{k\underline{\sigma}}_{r\underline{\rho}}(k(x,y))\,B^{r\underline{\rho}}_{i\underline{\mu}\, l\underline{\lambda}}. \tag{9.37}$$

Now suppose that (9.30) holds for some particular r. Then

$$\frac{\partial k^{i\underline{\mu}}(x,y)}{\partial y^{p_{r+1}\underline{\sigma}_{r+1}}\partial y^{p_r\underline{\sigma}_r}\dots\partial y^{p_1\underline{\sigma}_1}}$$

$$= \sum_{a=1}^{m+n}\sum_{\underline{\alpha}\in M_L}\sum_{s_1=1}^{m+n}\dots\sum_{s_r=1}^{m+n}\sum_{\underline{\tau}_1\in M_{L|s_1|},\underline{\gamma}_1\in M_L}\dots\sum_{\underline{\tau}_r\in M_{L|s_r|},\underline{\gamma}_r\in M_L}$$

$$T(r)^{a\underline{\gamma}_1\dots\underline{\gamma}_r}_{s_1\dots s_r}\,F^{\underline{\alpha}}_{\underline{\tau}_1\dots\underline{\tau}_r\underline{\gamma}_1\dots\underline{\gamma}_r}$$

$$\times\Bigg[\frac{1}{2}\,\tilde{\chi}^{b\underline{\beta}}_{p_{r+1}\underline{\sigma}_{r+1}}(x)\chi^{i\underline{\mu}}_{r\underline{\rho}}(k(x,y))B^{r\underline{\rho}}_{b\underline{\beta}\,a\underline{\alpha}}\tilde{\chi}^{s_1\underline{\tau}_1}_{p_1\underline{\sigma}_1}(y)\dots\tilde{\chi}^{s_r\underline{\tau}_r}_{p_r\underline{\sigma}_r}(y)$$

$$+\sum_{s=1}^r(-\frac{1}{2}\chi^{i\underline{\mu}}_{a\underline{\alpha}}(k(x,y))\,\tilde{\chi}^{s_1\underline{\tau}_1}_{p_1\underline{\sigma}_1}(y)\dots\tilde{\chi}^{s_{q-1}\underline{\tau}_{q-1}}_{p_{q-1}\underline{\sigma}_{q-1}}(y)B^{s_q\underline{\tau}_q}_{c\underline{\gamma}\,d\underline{\delta}}$$

$$\times\tilde{\chi}^{c\underline{\gamma}}_{p_{r+1}\underline{\sigma}_{r+1}}(y)\tilde{\chi}^{d\underline{\delta}}_{p_q\underline{\sigma}_q}(y)\tilde{\chi}^{s_{q+1}\underline{\tau}_{q+1}}_{p_{q+1}\underline{\sigma}_{q+1}}(y)\dots\tilde{\chi}^{s_r\underline{\tau}_r}_{p_r\underline{\sigma}_r}(y))\Bigg]. \tag{9.38}$$

Now the structure constants $C^k_{ij} = \sum_{\underline{\tau}\in M_L}C^{k\,\underline{\tau}}_{ij}\beta_{[\underline{\tau}]}$ of $\mathfrak{u}$ in the basis $\{X_i\}$ are related to the structure constants $B^{k\underline{\sigma}}_{i\underline{\mu}\, j\underline{\nu}}$ of $\mathfrak{h}$ in the basis $\{X_{i\underline{\mu}}\}$ by

$$B^{k\underline{\sigma}}_{i\underline{\mu}\, j\underline{\nu}} = \sum_{\underline{\tau}\in M_L}C^{k\,\underline{\tau}}_{ij}F^{\underline{\rho}}_{\underline{\nu}\underline{\mu}\underline{\tau}}. \tag{9.39}$$

Thus

$$\frac{\partial k^{i\underline{\mu}}(x,y)}{\partial y^{p_{r+1}\underline{\sigma}_{r+1}}\partial y^{p_r\underline{\sigma}_r}\dots\partial y^{p_1\underline{\sigma}_1}}$$

has the required form and so (9.30) is established by induction.

Further notation is now required. Given $\underline{\mu} \in M_L$ let $n(\underline{\mu})$ denote the number of sequences in M_L which contain $\underline{\mu}$ as a subsequence. Also, for $\underline{\nu}, \underline{\tau}$ and $\underline{\mu}$ in M_L define $A_{\underline{\mu}}^{\underline{\nu}\underline{\tau}}$ by

$$A_{\underline{\mu}}^{\underline{\nu}\underline{\tau}} = \frac{+1}{n(\underline{\tau})} \qquad \text{if } \beta_{[\underline{\nu}]}\beta_{[\underline{\tau}]} = \beta_{[\underline{\mu}]}$$

$$A_{\underline{\mu}}^{\underline{\nu}\underline{\tau}} = \frac{-1}{n(\underline{\tau})} \qquad \text{if } \beta_{[\underline{\nu}]}\beta_{[\underline{\tau}]} = -\beta_{[\underline{\mu}]}$$

$$A_{\underline{\mu}}^{\underline{\nu}\underline{\tau}} = 0 \qquad \text{otherwise.} \tag{9.40}$$

Using the summation convention, let

$$K_{p_1\ldots p_k}^i(x) = \left. \frac{\partial k^{i\underline{\mu}}(x,y)}{\partial y^{p_k\underline{\nu}_k}\ldots\partial y^{p_1\underline{\nu}_1}} \right|_{y=0}$$

$$\times A_{\underline{\mu}}^{\underline{\nu}_1\underline{\tau}_1} A_{\underline{\tau}_1}^{\underline{\nu}_2\underline{\tau}_2} \ldots A_{\underline{\tau}_{k-1}}^{\underline{\nu}_k\underline{\tau}_k} \beta_{[\underline{\tau}_k]} . \tag{9.41}$$

Now

$$\frac{\partial K^i(x,y)}{\partial y^{p_r\underline{\sigma}_r}\ldots\partial y^{p_1\underline{\sigma}_1}}(x,0)$$

$$= \sum_{a=1}^{m+n} \sum_{\underline{\alpha}\in M_L} \sum_{s_1=1}^{m+n} \ldots \sum_{s_r=1}^{m+n} \sum_{\underline{\tau}_1\in M_{L|s_1|}, \underline{\gamma}_1\in M_L} \ldots \sum_{\underline{\tau}_r\in M_{L|s_r|}, \underline{\gamma}_r\in M_L}$$

$$T(r)_{p_1\ldots p_r}^{a\underline{\gamma}_1\ldots\underline{\gamma}_r} F_{\underline{\sigma}_1\ldots\underline{\sigma}_r\underline{\gamma}_1\ldots\underline{\gamma}_r}^{\underline{\alpha}} \chi_{a\underline{\alpha}}^{i\underline{\mu}}(x)\beta_{[\underline{\mu}]} \tag{9.42}$$

since $\tilde{\chi}_{p\underline{\alpha}}^{s\underline{\tau}}(0) = \delta_p^s \delta_{\underline{\alpha}}^{\underline{\tau}}$. Thus when $x = 0$

$$\frac{\partial K^i}{\partial y^{p_1\underline{\sigma}_1}\ldots\partial y^{p_r\underline{\sigma}_r}}(x,0) = \beta_{[\underline{\sigma}_1]}\ldots\beta_{[\underline{\sigma}_r]}K_{p_1\ldots p_r}^i(x) . \tag{9.43}$$

Successive differentiation of (9.41) and (9.42) with respect to x, together with further applications of (9.36) and (9.37), shows that this result holds for all x. Thus, by Lemma 9.3.1, K is superanalytic in its first argument.

A parallel argument shows that K is also superanalytic in its second argument.

(c) By Corollary 9.3.2

$$\chi_j^i(x) = \partial_{j\,(y)}^S K^i(x,y)\Big|_{y=0}$$

$$= \frac{\partial k^{i\mu}}{\partial y^{j\underline{\nu}}}\Big|_{y=0} A_{\underline{\mu}}^{\nu\tau}\beta_{[\tau]}$$

$$= \chi_{j\underline{\nu}}^{i\mu}(x) A_{\underline{\mu}}^{\nu\tau}\beta_{[\tau]}, \tag{9.44}$$

so that

$$\partial_{k\underline{\rho}}\chi_j^i(0) = \partial_{k\underline{\rho}}\chi_{j\underline{\nu}}^{i\mu}(0) A_{\underline{\mu}}^{\nu\tau}\beta_{[\tau]}$$

$$= \tfrac{1}{2} C_{kj}^i \beta_{[\underline{\rho}]}. \tag{9.45}$$

Now χ_j^i is superanalytic (because it is a derivative of the superanalytic function K^i) and thus (again using Corollary 9.3.2)

$$\partial_i^S \chi_j^k(0) = \tfrac{1}{2} C_{ij\,[L]}^k. \tag{9.46}$$

∎

Recalling that the aim is to construct a super Lie group corresponding to a given super Lie module $\mathfrak{u}$, the previous proposition has established that a local super Lie group will certainly exist. It will now be shown that a topological group with a local superanalytic structure on a neighbourhood of the identity can be given a (unique) global superanalytic structure making it into a super Lie group. This is the analogue of a theorem familiar from classical Lie group theory.

Proposition 9.3.4 *Let G be a topological group and suppose that there is a neighbourhood V of the identity e of G on which is defined a function $\psi_e : V \to \mathbb{R}_{S[L]}^{m,n}$ which is a homeomorphism of V and its image (in the usual topology on a finite-dimensional vector space). Then, if the product in G is super analytic when expressed in terms of this chart, G can be defined in just one way as an (m,n)-dimensional super Lie group, such that the given chart, when restricted to a suitable nucleus, belongs to the superanalytic structure of G.*

Outline of proof Let W_1 and W_2 be neighbourhoods of e such that $W_1 W_2 \subset V$ and $W_1 W_2^{-1} \subset W_1$. Then the global superanalytic structure on G may be defined by the set of charts $\{(V_g, \psi_g) | g \in G\}$ with

$$V_g = W_2\, g \quad \text{and} \quad \psi_g(h) = \Psi_e(hg^{-1}). \tag{9.47}$$

The proof that this set of charts will give G the structure of a super Lie group is a straight forward generalisation of the classical proof.

The main theorem of this section can now be established.

Theorem 9.3.5   *Let* $\mathfrak{u}$ *be a* (m,n)-*dimensional super Lie module over* $\mathbb{R}_S$ *and* $L > n$ *be a positive integer. Also let* $\mathfrak{h}_{[L]}$ *be the even part of* $\mathfrak{u}_{[L]}$ *regarded as a* $2^{L-1}(m+n)$-*dimensional Lie algebra. If* $H_{[L]}$ *is a Lie group with Lie algebra* $\mathfrak{h}_{[L]}$, *then* $H_{[L]}$ *has the structure of an* (m,n)-*dimensional super Lie group over* $\mathbb{R}_{S[L]}$ *with super Lie module* $\mathfrak{u}_{[L]}$.

Proof   Lemma 9.3.3 and Proposition 9.3.4 together imply that $H_{[L]}$ can be given the structure of a super Lie group; also if X_i is the differential operator $\chi_i^j \partial_j^S$ then $\{X_i | i = 1, \ldots, m+n\}$ is a basis of the super Lie module of $\mathfrak{u}$ and if constants C'^k_{ij} are defined by

$$[X_i, X_j] = \sum_{k=1}^{m+n} C'^k_{ij} X_k \tag{9.48}$$

then (using the notation of Lemma 9.3.3)

$$C'^k_{ij} = \partial_i^S \chi_j^k(0) - (-1)^{|i||j|} \partial_j^S \chi_i^k(0). \tag{9.49}$$

Hence the super Lie module of $H_{[L]}$ is isomorphic to $\mathfrak{u}_{[L]}$. ∎

9.4   The super Lie groups which correspond to a given super Lie algebra

In this section the results of the previous section are exploited to show that, given a super Lie algebra $\mathfrak{g}$, a super Lie group H may be constructed with super Lie algebra $\mathfrak{g}$. This is achieved by showing that the universal covering group of the super Lie groups with super Lie module $\mathfrak{u} = \mathfrak{g} \otimes \mathbb{R}_S$ may be constructed as the direct product of a Lie group with Lie algebra $\mathfrak{g}_0$ and a nilpotent Lie group.

Theorem 9.4.1   *Suppose that* $\mathfrak{g} = \mathfrak{g}_0 + \mathfrak{g}_1$ *is an* (m,n)-*dimensional super Lie algebra. Let* $\mathfrak{u}_L = \mathbb{R}_{S[L]} \otimes \mathfrak{g}$ *and let* $\mathfrak{h}_L$ *denote the even part of* $\mathfrak{u}_L$ *regarded as a* $2^{L-1}(m+n)$-*dimensional Lie algebra. Then*

(a) $\mathfrak{h}_L \cong \mathfrak{g}_0 \oplus \mathfrak{n}_L$ *where* $\mathfrak{n}_L$ *is a nilpotent ideal in* $\mathfrak{h}_L$.

(b) If G *is the simply connected Lie group with Lie algebra* $\mathfrak{g}_0$ *and* N_L *is the simply connected Lie group with Lie algebra* $\mathfrak{n}_L$, *then a semi-direct*

product H_L of G and N_L can be defined which can be given the structure
of a super Lie group over $\mathbb{R}_{S[L]}$ with super Lie module $\mathfrak{u}_L$.

(c) For each $L > n$ there is a map $F_L : H_{[L]} \to H_{[L+1]}$, such that the direct
limit H of (H_L, F_L) has the structure of an (m,n)-dimensional super
Lie group over $\mathbb{R}_S$ with super Lie algebra $\mathfrak{g}$.

Proof   (a) Let $\{X_i | i = 1, \ldots, m+n\}$ be a super basis of $\mathfrak{u}$.   Then
$\left\{ X_{i\mu}\beta_{[\mu]} | i = 1, \ldots, m+n, \underline{\mu} \in M_{L\,|i|} \right\}$ is a basis of $\mathfrak{h}_L$.   Now the
span of $\{X_{i\emptyset} | i = 1, \ldots, m\}$ is a sub Lie algebra of $\mathfrak{h}_L$ which is iso-
morphic to $\mathfrak{g}_0$, while its complement $\mathfrak{n}_L$ in $\mathfrak{h}_L$ is the span of
$\left\{ X_{i\mu}\beta_{[\mu]} | i = 1, \ldots, m+n, \underline{\mu} \in M_{L\,|i|}, \underline{\mu} \neq \emptyset \right\}$ which forms a nilpotent ideal
in $\mathfrak{h}_L$.

(b) A representation $\mathrm{Adj}_\mathfrak{g}$ of $\mathfrak{g}_0$ on $\mathfrak{g}$ is defined by

$$\mathrm{Adj}_\mathfrak{g}(X)Y = [X,Y] \qquad X \in \mathfrak{g}_0, \quad Y \in \mathfrak{g}. \qquad (9.50)$$

Since G is simply connected, this representation can be exponentiated to
give a representation $\pi : G \to (\mathrm{aut}\,\mathfrak{g})_0$ (where $(\mathrm{aut}\,\mathfrak{g})$ denotes the set of
even vector space automorphisms of $\mathfrak{g}$).

Next define $\pi' : G \to \mathrm{aut}\,\mathfrak{h}_L$ by

$$\pi'(g)(aX) = a\pi(g)X \quad \forall a \in \mathbb{R}_{S[L]}, \quad X \in \mathfrak{g}_{|a|} \quad \text{and} \quad g \in G. \qquad (9.51)$$

Then π' maps $\mathfrak{n}_L$ into itself, so restricts to an automorphism $\pi'' = \pi'\big|_{\mathfrak{n}_L}$
of $\mathfrak{n}_L$. Finally, since the exponential map from $\mathfrak{n}_L$ to N_L is an analytic
diffeomorphism, the representation π'' can be exponentiated to give a representation $\alpha = \exp(\pi'')$ of G on N_L. Now H_L is defined to be the semi-
direct product $H_L = G \ltimes N_L$ with respect to the representation α, that is,
$H_L = G \times N_L$ with group operation $\circ$ defined by

$$(g_1, n_1) \circ (g_2, n_2) = (g_1 g_2, n_1\alpha(g_1)(n_2)). \qquad (9.52)$$

It will now be shown that H_L has the structure of a $2^{L-1}(m+n)$-dimensional
Lie group with Lie algebra $\mathfrak{h}_L$. First it is observed that N_L is homeomorphic
to $\mathbb{R}^{2^{L-1}(m+n)-m}$, since it is the exponentiation of a nilpotent Lie algebra
of this dimension; thus there is a global coordinate function $\eta : N_L \to \mathbb{R}^{2^{L-1}(m+n)-m}$ on N_L which is compatible with its Lie group structure.
Now let V, U, U_1 be connected coordinate neighbourhoods on the identity
e of G such that $UU \subset V$ and $UU_1^{-1} \subset U$. Let ϕ_e be an analytic chart

on U_1, canonical with respect to a basis $\{X_i | i = 1, \ldots, m\}$ of $\mathfrak{g}_0$. Then for each $g \in G$ define

$$U_g = U_1 g$$

$$\text{and} \quad \phi_g : U_g \to \mathbb{R}^m \qquad hg \mapsto \phi_e(h). \qquad (9.53)$$

Then the collection $\{(U_g, \phi_g) | g \in G\}$ forms an atlas on G compatible with the Lie group structure of G [32]. Because the representation α is smooth, the collection $\{(U_g \times N_L, \phi_g \times \eta) | g \in G\}$ is an atlas on $H_L = G \ltimes N_L$ which is compatible with its Lie group structure. The simplest way of establishing that the Lie algebra of H_L is $\mathfrak{h}_L$ is to consider elements near the identity in H_L. Let g_1 and g_2 be elements of G and n_1 and n_2 be elements of N_L, and express these elements in terms of the exponential maps from $\mathfrak{g}_0$ into G and $\mathfrak{n}_L$ into N_L respectively as

$$g_1 = \exp\left(\sum_{i=1}^{m} t x_1^{i\emptyset} X_i\right) , \quad g_2 = \exp\left(\sum_{i=1}^{m} t x_2^{i\emptyset} X_i\right) ,$$

$$n_1 = \exp\left(\sum_{i=1}^{m+n} \sum_{\underline{\mu} \in M_{L|i|}^*} t x_1^{i\underline{\mu}} X_{i\underline{\mu}}\right) \quad n_2 = \exp\left(\sum_{i=1}^{m+n} \sum_{\underline{\mu} \in M_{L|i|}^*} t x_2^{i\underline{\mu}} X_{i\underline{\mu}}\right)$$

$$(9.54)$$

with $x_1^{i\underline{\mu}}, x_2^{i\underline{\mu}}, t \in \mathbb{R}$ and M_L^* denoting the set of non-empty sequences in M_L. Then

$$\alpha(g_1) n_2 = \exp\left(\pi''(g_1) \left(\sum_{i=1}^{m+n} \sum_{\underline{\mu} \in M_{L|i|}^*} t x_2^{i\underline{\mu}} X_{i\underline{\mu}}\right)\right)$$

$$= \exp\left(t \sum_{i=1}^{m+n} \sum_{\underline{\mu} \in M_{L|i|}^*} x_2^{i\underline{\mu}} X_{i\underline{\mu}}\right.$$

$$+ t^2 \sum_{k=1}^{m} \sum_{i=1}^{m+n} \sum_{\underline{\mu} \in M_{L|i|}^*} x_1^{k\emptyset} x_2^{j\underline{\nu}} \left[X_{k\emptyset}, X_{j\underline{\nu}}\right] + O(t^3)\bigg) \quad (9.55)$$

and thus

$$(g_1, n_1) \circ (g_2, n_2)$$

$$= \left(\exp \left(t \sum_{i=1}^{m} (x_1^{i\emptyset} + x_2^{i\emptyset}) + \tfrac{1}{2}t^2 \sum_{i,j=1}^{m} x_1^{k\emptyset} x_2^{j\emptyset} \left[X_{k\emptyset}, X_{j\emptyset} \right] + O(t^3) \right), \right.$$

$$\times \exp \left(t \sum_{i=1}^{m+n} \sum_{\underline{\mu} \in M_{L|i|}^*} (x_1^{i\underline{\mu}} + x_2^{i\underline{\mu}}) X_{i\underline{\mu}} \right.$$

$$\left. \left. + \tfrac{1}{2}t^2 \sum_{i,j=1}^{m+n} \sum_{\underline{\mu} \in M_{L|i|}^*} \sum_{\underline{\nu} \in M_{L|j|}^*} x_1^{i\underline{\mu}} x_2^{j\underline{\nu}} \left[X_{i\underline{\mu}}, X_{j\underline{\nu}} \right] + O(t^3) \right) \right) \tag{9.56}$$

This establishes that the Lie algebra of H_L is $\mathfrak{h}_L$.

(c) The explicit construction of H_L shows that the natural inclusion ι_L of $\mathbb{R}_{S[L]}$ in $\mathbb{R}_{S[L+1]}$ induces homomorphisms $f_L : \mathfrak{h}_L \to \mathfrak{h}_{L+1}$ and $F_L : H_L \to H_{L+1}$ such that the direct limit $\mathfrak{h}$ of $(\mathfrak{h}_L, f_L)$ is a super Lie module which satisfies

$$\mathfrak{h} \cong (\mathbb{R}_S \otimes \mathfrak{g})_0 \tag{9.57}$$

and the direct limit H of (H_L, F_L) is a super Lie group with super Lie algebra $\mathfrak{g}$. ∎

Further super Lie groups with super Lie module $\mathbb{R}_S \otimes \mathfrak{g}$ may be constructed in a similar manner, using other Lie groups whose Lie algebra is $\mathfrak{g}_0$, provided that there are smooth representations of these Lie groups on $\mathfrak{g}$ which induce the representation Adj_g of $\mathfrak{g}_0$ on $\mathfrak{g}$. In the following section, where Kostant's construction of super Lie groups in the algebro-geometric framework is discussed, such representations play a central role, and are the key to the relationship between the two approaches to super Lie groups. Very similar arguments to those used in proving Theorem 9.4.1 lead to the following proposition.

Proposition 9.4.2 *Let G be Lie group with super Lie algebra $\mathfrak{g}$. Then $G_{[\emptyset]}$, the body of G, is a Lie group with group operation defined by $[g][g'] = [gg']$ where $[g]$ is the image of $g \in G$ under the projection $G \to G_{[\emptyset]}$. Also the Lie algebra of $G_{[\emptyset]}$ is $\mathfrak{g}_0$.*

The results of this section and the previous one reduces the classification theory of super Lie groups to that of super Lie modules.

9.5   Super Lie groups and the algebro-geometric approach to supermanifolds

In the algebro-geometric approach, a detailed development of the notion of super Lie group was given by Kostant [95], who used the terminology 'graded Lie group'. In this section the graded Lie groups of Kostant will be described, and related to the super Lie groups of Section 9.1.

An (m, n)-dimensional graded Lie group is an (m, n)-dimensional algebro-geometric supermanifold (G, A) with underlying manifold G an m-dimensional Lie group and sheaf A of super algebras over G having the product structure

$$A(U) \cong C^\infty(U) \otimes \Lambda(\mathbb{R}^n) \qquad (9.58)$$

for all open sets U in G, with further algebraic properties which will now be described. Associated to (G, A) is an (m, n)-dimensional super Lie algebra $\mathfrak{g} = \mathfrak{g}_0 \oplus \mathfrak{g}_1$ with $\mathfrak{g}_0$, the even part of $\mathfrak{g}$, isomorphic to the Lie algebra of G. Also, there exists a smooth representation $\pi : G \to (\operatorname{aut} \mathfrak{g})_0$ such that

$$\pi|_{G_e} = \operatorname{adj} \qquad (9.59)$$

where adj denotes the adjoint representation of G on $\mathfrak{g}$ obtained by exponentiating the action Adj of $\mathfrak{g}_0$ on $\mathfrak{g}$ defined by

$$\operatorname{Adj}_X Y = [X, Y] \quad \text{for all} \quad X \in \mathfrak{g}_0, Y \in \mathfrak{g}. \qquad (9.60)$$

This representation typifies the graded Lie group in the sense that two graded Lie groups (G, A) and (G, A') with the same underlying Lie group G and super Lie algebra $\mathfrak{g}$ are isomorphic if and only if A and A' are isomorphic sheaves and the representations of $\mathfrak{g}$ determined by A and A' are equivalent. As a consequence of this definition the sheaf A has a very interesting characterisation in terms of Lie Hopf algebras which is described in detail in [95].

In the remaining part of this section it is shown how Kostant's concept of graded Lie group relates to the super Lie groups of Section 9.1, more precisely, given a graded Lie group (G, A) with super Lie algebra $\mathfrak{g}$ it is shown that a (unique) super Lie group $\check{G}$ can be constructed whose body is G and whose super Lie module is $\mathbb{R}_S \otimes \mathfrak{g}$. As a supermanifold, the super Lie group $\check{G}$ is the supermanifold constructed from (G, A) according to the procedure developed in Section 8.4.

The first step is to construct an atlas of charts on G. Suppose that V, U and U_1 are connected coordinate neighbourhoods of the identity e of G such that $UU \subset V$ and $UU_1^{-1} \subset U$. Let $\phi_e : U_1 \to \mathbb{R}^m$ be an analytic coordinate map on U_1 which is canonical with respect to a basis $\{X_i | i = 1, \ldots, m\}$ of $\mathfrak{g}_0$. For each g in G define $U_g = U_1 g$ and coordinates on U_g by

$$\phi_g : U_g \to \mathbb{R}^m, \qquad \phi_g(h) = \phi_e(hg^{-1}). \tag{9.61}$$

Then the collection $\{(U_g, \phi_g) | g \in G\}$ is an analytic atlas of charts on G which is compatible with the Lie group structure [32]. The super Lie group $\check{G}$ will now be constructed by patching sets $S_g, g \in G$ with transition functions $\tau_{gg'}, g, g' \in G$, where

$$S_g = (\epsilon_{m,n})^{-1} (\phi_g(U_g))$$

and $\quad \tau_{gg'} : (\epsilon_{m,n})^{-1} (\phi_{g'}(U_g \cap U_{g'})) \to (\epsilon_{m,n})^{-1} (\phi_g(U_g \cap U_{g'}))$

with $\quad \tau^i = \widehat{\phi_g^i \circ \phi_{g'}^{-1}}, \qquad i = 1, \ldots, m$

and $\quad \tau^{j+m}(x; \xi) = \xi^j, \qquad j = 1, \ldots, n. \tag{9.62}$

Now let H be the super Lie group with super Lie module $\mathbb{R}_S \otimes \mathfrak{g}$ constructed from G and the representation of π by the method of Theorem 9.4.1. The next lemma shows that H (qua supermanifold) is superdiffeomorphic to the supermanifold $Y(G, A)$ constructed from the algebro-geometric superman-ifold (G, A) by the construction of Section 8.4.

Lemma 9.5.1 $Y(G, A)$ and H are superdiffeomorphic.

Proof This lemma is proved by showing that an atlas $\{(V_g, \psi_g) | g \in G\}$ can be defined on H such that $\psi_g(V_g) = S_g$ and $\psi_g \circ \psi_{g'}^{-1} = \tau_{gg'}$. Recall that ϕ_e is a coordinate map on a neighbourhood U_1 of the identity e of G, canonical with respect to the basis $\{X_1, \ldots, X_m\}$ of $\mathfrak{g}_0$. With the notation of Theorem 9.4.1, the analytic chart $(U_1 \times N, \phi_e \times \eta)$ on the neighbourhood $U_1 \times N$ of the identity of $H = G \ltimes N$ can be used to construct a superanalytic chart $(U_1 \times N, \psi_e)$ by the method of Lemma 9.3.3. Now let

$$V_g = U_g \times N$$

and $\quad \psi_g : V_g \to \mathbb{R}_{S[L]}^{m,n}$

with $\quad \psi_g(xg, n) = \psi_g((x, n) \cdot (g, e')) = \psi_e(x, n) \tag{9.63}$

where e' is the identity of N. Then the atlas $\{(V_g \psi_g) \, | g \in G\}$ defines the (unique) superanalytic structure on H with respect to which H is a super Lie group.

Also

$$\psi_g(V_g) = \iota^{-1} \left(\phi_g(U_g) \times \eta(N) \right)$$

$$= (\epsilon_{m,n})^{-1} \left(\phi_g(U_g) \right) . \tag{9.64}$$

Now $\psi_g \left(V_g \cap V_{g'} \right) \neq \emptyset$ if and only if $gg'^{-1} \in U$. Hence if $c = (a; \alpha)$ is an element of $\psi_g \left(V_g \cap V_{g'} \right)$ then $c = \psi_e(x, n)$ for some $x \in U_g$ and $n \in N$. Thus

$$\psi_g \circ \psi_{g'}^{-1}(c) = \psi_g \circ \psi_{g'}^{-1} \circ \psi_e(x, n)$$

$$= \psi_g(xg', n) . \tag{9.65}$$

Hence, for $1 = 1, \ldots, m$,

$$\psi_g^i \circ \psi_{g'}^{-1}(c) = \psi_g^i(xg', n)$$

$$= \psi_e^i(xg'g^{-1}, n)$$

$$= \phi_e^i(xg'g^{-1})1 + \sum_{\underline{\mu} \in M_{n,0}^*} \eta^{i\underline{\mu}}(n)\beta_{[\underline{\mu}]}$$

$$= \phi_g^i \circ \phi_{g'}^{-1} \circ \phi_e(x)1 + \sum_{\underline{\mu} \in M_{n,0}^*} \eta^{i\underline{\mu}}(n)\beta_{[\underline{\mu}]} . \tag{9.66}$$

Also for $j = 1, \ldots, n$

$$\psi_g^j \circ \psi_{g'}^{-1}(c) = \sum_{\underline{\nu} \in M_{n,1}} \eta^{j\underline{\nu}} \beta_{[\underline{\nu}]} . \tag{9.67}$$

However $\psi_g \circ \psi_{g'}^{-1}$ is a superanalytic function and so by Theorem 4.2.4 there exists a unique C^∞ functions $f^{k\underline{\mu}} : \phi_{g'}(U_g \cap U_{g'}) \to \mathbb{R}$, $k = 1, \ldots, m + n$, $\underline{\mu} \in M_{n,|k|}$ such that

$$\psi_g^k \circ \psi_{g'}^{-1}(c) = \sum_{\underline{\mu} \in M_{n,|k|}} \widehat{f^{k\underline{\mu}}}(a)\alpha_{\underline{\mu}} . \tag{9.68}$$

However

$$c = \psi_e(x, \eta)$$

$$= \left(\phi_e^1(x)1 + \sum_{\underline{\mu} \in M_{n,0}} \eta^{1\underline{\mu}}(n)\beta_{[\underline{\mu}]}, \dots, \phi_e^m(x)1 + \sum_{\underline{\mu} \in M_{n,0}} \eta^{m\underline{\mu}}(n)\beta_{[\underline{\mu}]}; \right.$$

$$\left. \sum_{\underline{\mu} \in M_{n,1}} \eta^{(m+1)\underline{\mu}}(n)\beta_{[\underline{\mu}]}, \dots, \sum_{\underline{\mu} \in M_{n,1}} \eta^{(m+n)\underline{\mu}}(n)\beta_{[\underline{\mu}]} \right), \qquad (9.69)$$

and comparison of (9.68) with (9.65) and (9.66) shows that for $i = 1, \dots, m$ and $j = 1, \dots, n$

$$f^{i\emptyset} = \phi_g^i \circ \phi_{g'}^{-1}, \qquad\qquad f^{i\underline{\mu}} = 0 \quad \text{if} \quad \underline{\mu} \neq \emptyset$$

$$f^{j\underline{\mu}} = 1 \qquad \text{if} \quad \underline{\mu} = (j), \qquad f^{j\underline{\mu}} = 0 \quad \text{otherwise}$$

so that

$$\psi_g^i \circ \psi_{g'}^{-1} = \widehat{\phi_g^i \circ \phi_{g'}^{-1}},$$

$$\text{and} \quad \psi_g^j \circ \psi_{g'}^{-1}(c) = c^{j+m}. \qquad (9.70)$$

∎

9.6 Super Lie group actions and the exponential map

In this section the exponential map of the even part of the super Lie module of a super Lie group is defined; this construction is very similar to the classical exponential map of the Lie algebra of a Lie group, and leads to one parameter subgroups also in close analogy with the classical case. The novel feature of the super setting is that the exponential map also leads to $(1,1)$ parameter sub super groups with elements of the form $\exp([\Xi, \Xi] t + \tau\Xi)$ where Ξ is an odd element of the super Lie module. These construction are used to show how a super Lie group action on a supermanifold induces a vector field on the manifold corresponding to each super Lie module element. Throughout this section G is a super Lie group with super Lie module $\mathfrak{u}$.

As in the classical case, the exponential map is constructed from the local one parameter group of local transformations corresponding to a given even vector field (c.f. Remark 6.5.7), in this case an even element of $\mathfrak{u}$.

Definition 9.6.1   Let $X \in \mathfrak{u}_0$. Suppose that $\phi : I_a \times U \to U$, where U is a neighbourhood of the identity e of G and $I_a = (-a, a), a > 0$ is a local 1 parameter group of local transformations induced by X. Let N be a positive integer such that $N > |1/a|$. Then the *exponential map* $\exp : \mathfrak{u}_0 \to G$ is defined by

$$\exp(X) = \left(\phi_{\frac{1}{N}} e\right)^N. \tag{9.71}$$

Very much as in the classical case, the exponential map can be shown to satisfy

$$\exp((s+t)X) = \exp(sX)\exp(tX) \tag{9.72}$$

for all $s, t \in \mathbb{R}$, so that (9.71) gives a definition of $\exp(X)$ which is independent of the choice of N and $\{\exp(tX) \,|\, t \in \mathbb{R}\}$ is a one parameter subgroup of G.

One might naively expect an odd analogue for an element Ξ of $\mathfrak{u}_1$, with subgroups consisting of elements $\exp(\Xi\tau)$ with τ an odd parameter. However, since in general $[\Xi, \Xi] \neq 0$, it is not in general true that $\exp(\sigma\Xi)\exp(\tau\Xi) = \exp((\sigma + \tau)\Xi)$, the true result being

$$\exp(\sigma\Xi)\exp(\tau\Xi) = \exp\left(-\tfrac{1}{2}\sigma\tau\,[\Xi, \Xi] + (\sigma + \tau)\Xi\right) \tag{9.73}$$

and there is in fact a $(1,1)$ parameter sub super group consisting of elements of the set $\left\{\exp(t\,[\Xi, \Xi] + \tau\Xi) \,|\, (t, \tau) \in \mathbb{R}_S^{1,1}\right\}$.

Proposition 9.6.2   *Let Ξ be an odd element of the super Lie module $\mathfrak{u}$ of the Lie group G. Suppose that $\Phi : (\epsilon_{1,1})^{-1}(I_a) \times U \to U$, where U is a neighbourhood of the identity e of G and $I_a = (-a, a), a > 0$, is a local $(1,1)$ parameter group of local transformations induced by Ξ. Let N be a positive integer such that $N > |t/a|$. Then a family of maps $\mathfrak{u}_1 \to G$ indexed by $\mathbb{R}_S^{1,1}$ is given by the prescription*

$$\exp(t\,[\Xi, \Xi] + \tau\Xi) = \left(\Phi_{\frac{t}{N}, \frac{\tau}{N}} e\right)^N. \tag{9.74}$$

These maps satisfy

$$\exp(t\,[\Xi, \Xi] + \tau\Xi)\exp(s\,[\Xi, \Xi] + \sigma\Xi)$$
$$= \exp((t + s + \sigma\tau)\,[\Xi, \Xi] + (\tau + \sigma)\Xi) \tag{9.75}$$

for all $(s, \sigma), (t, \tau)$ in $\mathbb{R}_S^{1,1}$.

This result follows directly from Theorem 6.5.9.

Finally the use of the exponential map to define a vector field on a space where G acts corresponding to an element of the super Lie module $\mathfrak{u}$ of G is described. Suppose that E is a right G-space, that is, it is a supermanifold which carries a supersmooth right G action $(p, g) \mapsto pg$, $g \in G, p \in E$. In the case of even elements of $\mathfrak{u}$, the construction is essentially the same as the classical one, while odd elements require the $(1, 1)$ parameter subgroups following from the preceding proposition.

Definition 9.6.3

(a) If $Y \in \mathfrak{u}_0$, the fundamental vector field on E corresponding to Y is defined by

$$\underline{Y}_p f = \frac{df}{dt} (p \exp{(tY)}) \bigg|_{t=0} \tag{9.76}$$

for $p \in E$, $f \in G^\infty(E)$; also

(b) if and $\Xi \in \mathfrak{u}_1$ the fundamental vector field on E corresponding to Ξ is defined by

$$\underline{\Xi}_p f = \left(\frac{\partial^O}{\partial \tau} + \tau \frac{\partial^E}{\partial t} \right) f(p \exp{(t\,[\Xi, \Xi] + \tau\Xi)}) \bigg|_{(t,\tau)=(0,0)} \tag{9.77}$$

for $p \in E$, $f \in G^\infty(E)$.

Chapter 10

Tensors and forms

As in classical differential geometry, there are two equivalent ways of defining a tensor on a supermanifold. One is to use the bundle of super frames, described in Section 12.2, defining various classes of tensors on an (m, n)-dimensional supermanifold as cross-sections of the associated super bundle of the bundle of super frames corresponding to representations of its structure group $\mathrm{GL}(m, n; \mathbb{R})$. The other approach, which is briefly described in Section 10.1, uses the tensor algebra of the vector fields on the supermanifold. General tensors are described and then in Section 10.2 the notion of Berezin density, which generalises the standard notion of density, is introduced. The extension of differential forms to the super setting can be made in more than one way. Section 10.3 considers exterior forms, which are a graded analogue of classical differential forms, while Section 10.4 introduces mixed covariant and contravariant tensors, referred to as super forms, which were introduced by Voronov and Zorich [156] and are important in integration theory.

Throughout this chapter U is an open subset of an (m, n)-dimensional supermanifold $\mathcal{M}$. The constructions can be understood in either the concrete or the algebro-geometric language; in general they are presented in G^∞ form, restrictions to H^∞ are normally straightforward.

10.1 Tensors

The concept of vector field was introduced in Chapter 6, and forms the basis of the construction of tensors which will now be described. A tensor on U is an element of the tensor algebra of $\mathcal{D}(U)$; this tensor algebra has three degrees, the usual $\mathbb{Z} \oplus \mathbb{Z}$ degree corresponding to the contravariant and covariant degree and the $\mathbb{Z}_2$ degree which arises in the super setting.

The definition of an r-covariant tensor on U will be given explicitly.

Definition 10.1.1   An *r-covariant tensor* on U is a mapping

$$\alpha : (\mathcal{D}(U))^r \to G^\infty(U), \ (X_1, \ldots, X_r) \mapsto \langle X_1, \ldots, X_r | \alpha \rangle$$

which is $G^\infty(U)$ multilinear in the sense that

$$\langle X_1, \ldots, fX_i, \ldots, X_r | \alpha \rangle = (-1)^{|f|(\sum_{l=1}^{i-1} |X_l|)} f \langle X_1, \ldots, X_r | \alpha \rangle \quad (10.1)$$

for all f in $G^\infty(U)$ and $i = 1, \ldots, r$. The set of r-covariant tensors on U is denoted $T^r(U)$.

The space of covariant tensors has a natural right $G^\infty(U)$ module structure given by

$$\langle X_1, \ldots, fX_i, \ldots, X_r | \alpha f \rangle = \langle X_1, \ldots, fX_i, \ldots, X_r | \alpha \rangle f . \quad (10.2)$$

A Z_2 degree can be put on the space of covariant tensors in the standard way, that is, $|\alpha| = i$ if

$$| \langle X_1, \ldots, X_r | \alpha \rangle | = \sum_{i=1}^{r} |X_i| + i \quad \text{modulo} \quad 2 \quad (10.3)$$

for all $X1, \ldots, X_r$ in $\mathcal{D}(U)$.

One-covariant tensors are as in the classical case known as one-forms; the space of one forms is denoted $\Omega^1(U)$. If U is a coordinate neighbourhood with local coordinates $(x^1, \ldots, x^m; \xi^1, \ldots, \xi^n)$ then $\Omega^1(U)$ is a free $G^\infty(U)$ module with super basis $\{dx^1, \ldots, dx^m; d\xi^1, \ldots, d\xi^n\}$ where

$$\left\langle \frac{\partial}{\partial x^i} \Big| dx^j \right\rangle = \delta_i^j, \ \left\langle \frac{\partial}{\partial x^i} \Big| d\xi^j \right\rangle = 0, \ \left\langle \frac{\partial}{\partial \xi^i} \Big| dx^j \right\rangle = 0, \ \left\langle \frac{\partial}{\partial \xi^i} \Big| d\xi^j \right\rangle = \delta_i^j . \quad (10.4)$$

Contravariant and mixed tensors can also be defined.

10.2   Berezinian densities

On a classical manifold the notion of tensor density can be useful in the context of integration. For an (m, n)-dimensional supermanifold the analogous object is one with a single component in a given coordinate system which transforms according to the superdeterminant of the $(m, n) \times (m, n)$

matrix

$$\begin{pmatrix} \partial x_\alpha^i / \partial x_\beta^j & \partial x_\alpha^i / \partial \xi_\beta^j \\ \partial \xi_\alpha^i / \partial x_\beta^j & \partial \xi_\alpha^i / \partial \xi_\beta^j \end{pmatrix} \tag{10.5}$$

whose entries are derivatives of the coordinate transition functions from a chart (V_α, ψ_α) to a chart (V_β, ψ_β) (with corresponding local coordinates $(x_\alpha; \xi_\alpha)$ and $(x_\beta; \xi_\beta)$ respectively). This superdeterminant, which is the super analogue of the Jacobian, is known as the *Berezinian* and will be written $\mathrm{Ber}\left(\frac{x_\alpha, \xi_\alpha}{x_\beta, \xi_\beta}\right)$ or $\mathrm{Ber}\left(\psi_\alpha \circ \psi_\beta^{-1}\right)$.

Definition 10.2.1   A *Berezinian density* ω is an object which in each local coordinate system (V_α, ψ_α) has local representative $\omega_\alpha(x_\alpha; \xi_\alpha)$, with local representatives for (V_α, ψ_α) and (V_β, ψ_β) related by

$$\omega_\beta(x_\alpha; \xi_\alpha) = \omega_\alpha(x_\alpha; \xi_\alpha) \mathrm{Ber}\left(\frac{x_\alpha, \xi_\alpha}{x_\beta, \xi_\beta}\right) \tag{10.6}$$

on $V_\alpha \cap V_\beta$.

A more complete and formal definition of Berezinian density is given in Definition 12.2.2. Examples may be found in (12.14) and Section 13.2. The Berezin densities on a supermanifold define a sheaf over $\mathcal{M}_{[\emptyset]}$ which is denoted $\mathfrak{Ber}$.

10.3   Exterior differential forms

In this section the graded analogue of differential forms are constructed largely following the work of Kostant [95]; the exterior product and exterior derivative are defined. Such forms are not used in the standard integration theory on supermanifolds, and do not (in the smooth setting) have de Rham cohomology which encodes any more topological information than that of the underlying manifold. However they have been applied in an interesting and powerful way by Howe, Raetzel and Sezgin in [77] to construct actions for super membranes using the superembedding technique, as is described in Section 13.3. The tensors constructed in this section are covariant graded antisymmetric tensors, and will be referred to as exterior forms when necessary to distinguish them from the super forms introduced in the following section, which have mixed covariant and contravariant properties and do relate to the standard integration theory on a supermanifold.

The basic concept is a p-form, which will now be defined.

Definition 10.3.1   Let p be a positive integer. A *p-form* on U is a p-covariant tensor α on U which is graded antisymmetric, that is, for $X_1, \ldots, X_p$ in $\mathcal{D}(U)$ and $i = 1, \ldots, p-1$

$$\langle X_1, \ldots, X_{i-1}, X_{i+1}, X_i, \ldots X_p | \, \alpha \rangle$$
$$= (-1)^{|X_i||X_{i-1}|+1} \langle X_1, \ldots, X_{i-1}, X_i, X_{i+1}, \ldots X_p | \, \alpha \rangle \, . \quad (10.7)$$

The space of p forms on U is denoted $\Omega^p(U)$.

This definition is consistent, in the case where $p = 1$, with the concept of one-form introduced in the previous section. Because these tensors satisfy a graded antisymmetry condition, there is no longer any upper limit on the degree of a form and the full space of forms can be defined to be

$$\Omega(U) = \bigoplus_{p=0}^{\infty} \Omega^p(U) \quad (10.8)$$

with $\Omega^0(U) = G^\infty(U)$. This space is bigraded, with a $\mathbb{Z}_2$ degree coming from the super algebra aspect and a $\mathbb{Z}$ degree from the form. An operation of exterior multiplication will be defined which gives $\Omega(U)$ the structure of a bigraded commutative algebra, and an operation of exterior differentiation will be defined which is a bigraded odd super derivation on this algebra.

Definition 10.3.2   Let p and q be positive integers and α be a p-form and β a q-form on U. Then $\alpha \wedge \beta$ is defined to be the mapping

$$\alpha \wedge \beta : (\mathcal{D}(U))^{p+q} \to G^\infty U$$
$$(X_1, \ldots, X_{p+q}) \mapsto \sum_{\lambda \in M_{p+q}} (-1)^{\left(\sigma\left(X_\lambda, X_{\lambda^C}\right) + |\beta||X_\lambda|\right)} \langle X_\lambda | \, \alpha \rangle \langle X_{\lambda^C} | \, \beta \rangle$$

$$(10.9)$$

where if $\lambda = (\lambda_1, \ldots, \lambda_k)$ is a multi index in M_{p+q} then λ^C is the complementary multi index, $X_\lambda = (X_{\lambda_1}, \ldots, X_{\lambda_k})$ and $|X_\lambda| = \sum_{r=1}^{k} |X_{\lambda_r}|$. Also $\sigma(X_\lambda, X_{\lambda^C}) = \sum(1 + |X_r||X_{r'}|)$ where the sum is over all pairs (r, r') in $\lambda \times \lambda'$ where $r > r'$.

This product can be extended to a product on all of $\Omega(U)$ by defining multiplication by zero forms to be the right $G^\infty(U)$ module action (10.3). The key algebraic properties of this product are gathered in the following proposition. The proof is omitted since it is a reasonably straightforward generalisation of the standard case. Details may be found in [95].

Proposition 10.3.3   *With the notation of Definition 10.3.2* *(a)* $\alpha \wedge \beta$ *is a $p+q$-form and* $|\alpha \wedge \beta| = |\alpha| + |\beta|$.

(b) The exterior product is associative.

(c) The exterior product is bigraded commutative, that is

$$\alpha \wedge \beta = (-1)^{(pq+|\alpha||\beta|)}\beta \wedge \alpha. \tag{10.10}$$

If U is a coordinate neighbourhood then a p-form α can be expressed as

$$\alpha = \sum_{\substack{j_1=1\ldots j_p=1 \\ j_1 \leq \cdots \leq j_p}}^{p} \mathrm{d}X^{j_1} \wedge \cdots \wedge \mathrm{d}X^{j_p}\alpha_{j_1 \ldots j_p} \tag{10.11}$$

where $\alpha_{j_1 \ldots j_p} \in G^{\infty}(U)$ and $\mathrm{d}X^i$ are as defined in (10.4).

The exterior derivative can now be defined by the following theorem, which establishes both its existence and its properties. Again the proof is omitted because of similarities with the proof of the analogous classical result. There are many different routes to this construction, here that of Sternberg [147] is followed.

Theorem 10.3.4 *There exists a uniquely defined map* $\mathrm{d} : \Omega(U) \to \Omega(U)$ *such that*

(a) $\mathrm{d}(\alpha_1 + \alpha_2) = \mathrm{d}\alpha_1 + \mathrm{d}\alpha_2$

(b) $\mathrm{d}(\alpha \wedge \beta) = \mathrm{d}\alpha \wedge \beta + (-1)^p\alpha \wedge \mathrm{d}\beta$ *if* $\alpha \in \omega^p(\mathcal{M})$

(c) If f is a zero-form, $\mathrm{d}f$ has a well-defined expression in local coordinates

$$\mathrm{d}f = \sum_{i=1}^{m}\mathrm{d}x^i\frac{\partial f}{\partial x^i} + \sum_{j=1}^{n}\mathrm{d}\xi^j\frac{\partial f}{\partial \xi^j} \tag{10.12}$$

(d) $\mathrm{d}(\mathrm{d}f) = 0$.

This definition immediately shows that if U an open and connected subset of $\mathcal{M}$ and $f \in G^{\infty}(U)$ then $\mathrm{d}f = 0$ if and only if f is constant.

On a classical manifolds differential forms encode a considerable amount of information about global features of the manifold. Much of this information comes via integration and Stokes' theorem. There is no direct analogy of this situation on a supermanifold, for a number of reasons. To begin with, the theory of integration is rather different. It is possible to integrate p forms over p chains almost exactly as in the classical setting (with p chains defined in terms of maps of simplices in $\mathbb{R}^p$ into the supermanifold) but the results do not seem to convey global information beyond that relating to the body. (However such integrals do have applications to superembeddings, as is explained in Section 13.3, and so they are defined in Section 11.7.) Related to this is the fact that the de Rham cohomology groups of a smooth supermanifold are simply those of the underlying manifold, as will be demonstrated below. A notion of super contour integration has been introduced in the context of super Riemann surfaces; this is described in Chapter 14, but the objects which are integrated are not exterior forms, instead they are examples of the super forms described in the following section.

Given that the data of a smooth supermanifold is that of a vector bundle over the underlying manifold, the following de Rham theorem, due originally to Kostant [95], is perhaps not surprising.

Theorem 10.3.5   *The de Rham cohomology of H^∞ forms on the supermanifold $\mathcal{M}$ is isomorphic to that of the underlying manifold $\mathcal{M}_{[\emptyset]}$.*

(The restriction to H^∞ is for simplicity. Because of Batchelor's theorem 8.2.1 the effect of extending to G^∞ would simply be a tensor product with $\mathbb{R}_S$.) This result is proved in two stages; first the Poincaré lemma is proved for a supermanifold.

Lemma 10.3.6   *Suppose that U is a contractible coordinate neighbourhood of $\mathcal{M}$ and that $\alpha \in \Omega^p(\mathcal{M})$ (with $p \geq 1$) satisfies $\mathrm{d}\alpha = 0$. Then there exists $\beta \in \Omega^{p-1}(\mathcal{M})$ such that $\mathrm{d}\beta = \alpha$.*

Outline of proof   The space $\Omega(U)$ may be decomposed as

$$\Omega(U) = C \otimes D \tag{10.13}$$

where $C \equiv \Omega(U_{[\emptyset]}) \otimes \mathbb{R}_S$ and $D \equiv \Lambda(\mathbb{R}^n) \otimes S(\mathbb{R}^n)$, the decomposition corresponding to expression of a form in $\Omega^p(\mathcal{M})$ as a sum of terms of the form

$$f_{i_1 \ldots i_l}(x)\mathrm{d}x^{i_1} \ldots \mathrm{d}x^{i_l} \otimes g_{j_{l+1} \ldots j_p}(\xi)\mathrm{d}\xi^{i_{l+1}} \ldots \mathrm{d}\xi^{i_p} \tag{10.14}$$

with d acting as $\mathrm{d}_{[\emptyset]} \otimes 1 \oplus 1 \otimes \mathrm{d}_k$ where $\mathrm{d}_{[\emptyset]}$ denotes the exterior derivative on $\Omega(U_{[\emptyset]})$ and d_k denotes the standard derivative of the Koszul complex $\Lambda(\mathbb{R}^n) \otimes S(\mathbb{R}^n)$. Since this complex is acyclic and $\Omega(U_{[\emptyset]})$ satisfies the Poincaré lemma, the result is proved.   ∎

Proof of Theorem 10.3.5 Using a partition of unity as in Theorem 4.8.1 shows that the sheaf $U_{[\emptyset]} \to \Omega^p(U)$ is flasque (or flabby); thus the exterior derivative defines

$$\cdots \to \Omega^p(U) \to \Omega^{p+1}(U) \to \ldots, \tag{10.15}$$

a flasque resolution of the constant sheaf, so that the de Rham cohomology of $\Omega(\mathcal{M})$ is isomorphic to the Čech cohomology of $\mathcal{M}_{[\emptyset]}$ with coefficients in $\mathbb{R}_S$.

10.4 Super forms

A super form is an object which has a double degree, being partly covariant (and graded antisymmetric) and partly contravariant (and graded symmetric). The motivation for considering these objects arises from the desire to find the appropriate notion of 'top form' on a supermanifold, in particular an object whose single component will transform by the Berezinian or superdeterminant under change of basis of the tangent space. The simplest notion might appear to be to define an (m, n) super form on an (m, n)-dimensional supermanifold to be an object V written in local coordinates as

$$V = V(x; \xi) \, \mathrm{d}x^1 \wedge \cdots \wedge \mathrm{d}x^m \otimes \frac{\partial}{\partial \xi^1} \wedge \cdots \wedge \frac{\partial}{\partial \xi^j} \qquad (10.16)$$

declaring that the component $V(x; \xi)$ transform by the Berezinian. This of course leads to the notion of Berezinian density as defined in Section 10.2, which, while useful in some contexts, does not provide an entirely ideal or complete integration theory, as will emerge in Chapter 11. It does however provide a reasonably workable theory in many applications, for instance in defining actions of supersymmetric theories; examples of this are given in Section 13.2. However for a more complete theory of integration a more general notion of (r, s) super form on an (m, n)-dimensional manifold for any (r, s) with $r < m$ and $s < n$ is required. Unfortunately there does not seem to be any construction that allows a full analogue of conventional integration together with its geometric and topological applications, although many features can be achieved, and there are a number of more or less equivalent versions with varying notation and terminology.

The naive definition of (r, s) super form is a (r, s) tensor which is graded antisymmetric in the r covariant slots and graded symmetric in the s contravariant slots; the difficulty is using such an object in theory of integration is that there is not always a well defined pullback of such an object from the supermanifold where it resides to a standard super simplex. In Section 11.4 one alternative notion of super form, due to Rothstein [133], where the contravariant part of the tensor is replaced by a differential operator, is described and related to the Berezin integral.

An elegant and powerful notion of super form, which lead to a considerable advance in the geometric theory of integration on supermanifolds, was introduced by Voronov and Zorich [156] and is further developed, and described in detail, in the book of Voronov [153]. The theory of super forms leads to a notion of (r, s) super form on an (m, n)-dimensional manifold for any values of r and s with $r < m$ and $s < n$. In seeking to construct a super form, an approach to classical differential forms must be found which can be generalised to the super setting in an appropriate way. The insight of Voronov and Zorich was that, rather than using representations of $\mathrm{GL}(m, n)$ (generalising the use of representations of $\mathrm{GL}(m)$ for classical forms), to recognise that the important feature of a p-form is that it can be integrated over a p-dimensional surface, and that this aspect was the appropriate characteristic to generalise. Thus the notion of super form involves an integrand or (in the terminology of [153]) a Lagrangian.

Suppose that $\sigma : I^p \to M$ is a smooth and proper map of the unit p-cube I^p into an m-dimensional manifold M, so that it defines a parameterised p-dimensional surface in M. (The unit p-cube is the set $\{t = (t^1, \ldots, t^p) | 0 \le t^a \le 1, a = 1, \ldots, p\} \subset \mathbb{R}^p$.) Then, if ω is a p-form on M, the pull-back $\sigma^*\omega$ has the form

$$\sigma^*\omega(t) = \mathcal{L}\left(x^a(t), \dot{x}^i_a(t)\right)\, \mathrm{d}t^1 \wedge \cdots \wedge \mathrm{d}t^p \qquad (10.17)$$

at a point $t \in I^p$ where $\sigma(t) = (x^1(t), \ldots, x^m(t))$ in some local coordinate system x on M and $\dot{x}^i_a = \frac{\partial x^i}{\partial t^a}$. The function $\mathcal{L}$, whose domain is a subset of $\mathbb{R}^{m+mp}$, satisfies

$$\frac{\partial^2 \mathcal{L}}{\partial \dot{x}^i_a \partial \dot{x}^j_b} + \frac{\partial^2 \mathcal{L}}{\partial \dot{x}^i_b \partial \dot{x}^j_a} = 0\,. \qquad (10.18)$$

$$\mathcal{L}\left(x^i, \sum_{b=1}^{p} g^b_a \dot{x}^i_b\right) = \det(g^b_a)\mathcal{L}(x^i, x^i_a)\,. \qquad (10.19)$$

These two properties are the aspects of the 'Lagrangian' obtained from the p-form ω which are generalised to give the concept of super form. In the classical setting, these properties hold not simply for pull back of forms on M, but also for more general objects such as densities on p-dimensional surfaces in M. These ideas lead to the notion of super form which will now be developed.

The full definition of a super form requires the notion of jet bundle. It will be sufficient to consider only trivial bundles, but bundle terminology is used since this is conventional in this context. For a slightly more informal definition the reader may simply use (10.25) and the sentence which follows.

Definition 10.4.1

(a) The *unit (r,s)-cube* $I^{r,s}$ is the set $(\epsilon_{r,s})^{-1}(I^r)$. A typical point is denoted

$$T = (t; \tau) = (t^1, \ldots, t^r; \tau^1, \tau^s) = (T^1, \ldots, T^{r+s}). \qquad (10.20)$$

(b) Consider the trivial bundle $I^{r,s} \times \mathcal{M}$ over $I^{r,s}$ where $\mathcal{M}$ is an (m,n)-dimensional supermanifold with $m \geq r$ and $n \geq s$.

   (i) Two sections f and g of $I^{r,s} \times \mathcal{M}$ are said to have *first order contact* at $T \in I^{r,s}$ if they have the same value, that is $f(T) = g(T)$, and for all tangent vectors Y in $T_T I^{r,s}$

$$f_* Y_{f(T)} = g_* Y_{g(T)}. \qquad (10.21)$$

   (ii) The equivalence class containing f is denoted $j^1_T f$ and is called the *first jet* of f. The set of all equivalence classes of sections with value u at T is denoted $J^1_{T,u}(I^{r,s} \times \mathcal{M})$.

   (iii) The total space of the first jet bundle of $I^{r,s} \times \mathcal{M}$ is

$$J^1(I^{r,s} \times \mathcal{M}) = \sqcup_{(T,u) \in I^{r,s} \times \mathcal{M}} J^1_{T,u}(I^{r,s} \times \mathcal{M}). \qquad (10.22)$$

(c) A jet f is said to be *proper* if in any coordinate system the rank of the $n \times s$ matrix $\left(\frac{\partial f^{j+m}}{\partial \tau^a} \big| j = 1, \ldots n, a = 1, \ldots s \right)$ is s. (Here local coordinates on $\mathcal{M}$ are used to define the components of f, and a cross section of the trivial bundle $I^{r,s} \times \mathcal{M}$ is simply regarded as a mapping from $I^{r,s}$ into $\mathcal{M}$.) The total space of the first proper jet bundle of $I^{r,s} \times \mathcal{M}$ is denoted $Jp^1(I^{r,s} \times \mathcal{M})$.

The first jet bundle is a supermanifold of dimension $(m + mr + ns, n + ms + nr)$. It has local coordinates

$$(X^i, \dot{X}^i_a | i = 1, \ldots, m + n, a = 1, \ldots, r + s) \qquad (10.23)$$

corresponding to local coordinates X^i on $\mathcal{M}$ and T^a on $I^{r,s}$. (In the language of classical mechanics, the jet bundle is the configuration space, with X being position and $\dot{X}$ velocity.)

Now, generalising the features of the pull-back of a classical from described above in such a way that the role of the Jacobian is played by the Berezinian, a super form can be defined in the following way.

Definition 10.4.2 Let U be an open set in $\mathcal{M}$. Then an (r, s) *super form* on U is an equivalence class of triples $(\mathcal{L}_\alpha, \psi_\alpha)$ where ψ_α is a coordinate system on $I^{r,s}$ and $\mathcal{L}_\alpha$ is a supersmooth mapping

$$\mathcal{L}_\alpha : Jp^1(I^{r,s} \times U) \to \mathbb{R}_S \qquad (10.24)$$

and $(\mathcal{L}_\alpha, \psi_\alpha) \sim (\mathcal{L}_\beta, \psi_\beta)$ if and only if

$$\mathcal{L}_\beta(X^i, \dot{X}^i_{a_\beta}) = \mathrm{Ber}\left(\psi_\beta \circ \psi_\alpha^{-1}\right) \mathcal{L}_\alpha(X^i, \dot{X}^i_{a_\alpha}). \qquad (10.25)$$

In local coordinates a super form is thus represented by a single component $\mathcal{L}\left(X^i, \dot{X}^i_a\right)$ which transforms according to the Berezinian of changes of coordinate on $I^{r,s}$. In Chapter 11 the integral of a super form will be constructed.

Super forms pull back under mappings between supermanifolds, because jets push forward: suppose that if $h : \mathcal{M} \to \mathcal{N}$ is a supersmooth mapping of supermanifolds and f and g are two mappings of $I^{r,s}$ into $\mathcal{M}$ which have the same one jet on $I^{r,s} \times \mathcal{M}$. Then $h \circ f$ and $h \circ g$ have the same one jet on $I^{r,s} \times \mathcal{N}$. Thus the push-forward of $j^1 f$ by h can be defined by

$$h_*(j^1 f) = j^1(h \circ f). \qquad (10.26)$$

This leads to a definition of pull-back of a super form on $\mathcal{N}$ to a super form on $\mathcal{M}$. This is defined using local coordinates on $I^{r,s}$, so that the pull back of $\mathcal{L}_\alpha$ via h is $h^* \mathcal{L}_\alpha$ where

$$h^* \mathcal{L}_\alpha(j) = \mathcal{L}_\alpha(h_* j). \qquad (10.27)$$

The exterior derivative of the (r, s)-form $\mathcal{L}$ is the $(r + 1, s)$-form defined by

$$\mathrm{d}\mathcal{L}_\alpha = (-1)^r \dot{X}^i_{r+1_\alpha} \frac{\partial}{\partial \mathcal{L}_\alpha} \dot{X}^i_{a_\alpha}(X^i, \dot{X}^i_{a_\alpha}). \qquad (10.28)$$

In Chapter 11 the notion of super form will allow a geometric integration theory for supermanifolds which contains some of the features of geometric integration on manifolds.

Chapter 11

Integration on supermanifolds

It is with integration that the differences between conventional analysis and superanalysis become most pronounced. There are various features of functions of anticommuting variables which suggest that the extension of classical integration theory to the super setting will not be straightforward. One difference arises when considering integration as the inverse of differentiation, because there exist functions of a single anticommuting variable which are not the derivative of any function. (An example of such a function is $f(\xi) = \xi$.) Another difference is that it is the superdeterminant defined in Section 2.3 rather than a determinant that has the multiplicative property; this suggests the use of objects which transform under change of coordinate by the superdeterminant of the matrix of coordinate derivatives (usually referred to as a the Berezinian), objects which are not simply exterior differential forms.

The starting point is the Berezin integral over anticommuting variables, defined in the first section of this chapter. This integral is algebraic, and does not appear to have the elementary properties of an integral. It is not an antiderivative (indeed it more closely resembles a derivative) and it has no measure-theoretic interpretation, perhaps not surprisingly in view of the DeWitt topology with its all or nothing approach to the odd sector. The justification for the definition is its usefulness – which appears in many areas – and its ability to combine with standard integrals of commuting variables to give an integral whose transformation properties involves the Berezinian superdeterminant, leading to an elegant theory of integration at least on compact supermanifolds.

The first section of this chapter considers integration on the purely odd superspace $\mathbb{R}_S^{0,n}$, while in Section 11.2 this integration is combined with standard integration on $\mathbb{R}^m$ to give a theory of integration on $\mathbb{R}_S^{m,n}$, including a transformation rule for G^∞ change of coordinate. This transformation rule is valid only for functions which have compact support or tend to zero sufficiently rapidly on the boundaries of the region of integration. In the following section the theory of integration over compact supermanifolds, which has been known in a complete and consistent form for some years (see [99] and references therein) is described. This theory makes use of the rule for change of variable for integrals on $\mathbb{R}_S^{m,n}$.

Various versions of integration using contour integrals for the even part have been developed to cover the situation where the integrand does not vanish on the boundary of the region concerned [43, 121, 151], but the approaches adopted by Rothstein and Voronov seem more effective. Section 11.4 contains an account of Rothstein's approach to integration on non-compact supermanifolds. Voronov's geometric integration theory, which uses the super forms constructed in Section 10.4, is described in Section 11.5.

An exception to the general outlook of integration on supermanifolds occurs in $(1,1)$ dimensions, where integration on supercurves has analogues of both the measure-theoretic and antiderivative aspects of conventional integrals, as is described in section Section 11.6. In Chapter 14 this leads to a theory of contour integration on super Riemann surfaces. One may speculate that higher dimensional analogues of this approach would lead to a full geometric theory of integration on supermanifolds.

Finally, a theory of integration of exterior forms is presented. This is very simple, but finds useful application in the superembeddings approach to supersymmetric extended objects discussed in Section 13.3.

11.1 Integration with respect to anti commuting variables

The theory of integration on the purely odd superspace $\mathbb{R}_S^{0,n}$ is based on the following definition:

Definition 11.1.1   Suppose that f is a G^∞ function of $\mathbb{R}_S^{0,n}$ into $\mathbb{R}_S$ with

$$f(\xi^1, \ldots, \xi^n) = f_{1\ldots n}\xi^1 \ldots \xi^n + \text{ lower order terms.} \qquad (11.1)$$

Then the *Berezin integral* of f is defined to be

$$\int \mathrm{d}^n\xi \, f(\xi^1, \ldots, \xi^n) = f_{1\ldots n}. \qquad (11.2)$$

Applications of the Berezin integral include the construction of invariants for supersymmetric theories (c.f. Sections 13.1 and 13.2), where it is important that G^∞ functions are used rather than simply H^∞ functions. It is possible to extend the Berezin integral to include functions f whose expansion coefficients f_μ take values in a more general algebra than $\mathbb{R}_S$. It is hard to give an *a priori* motivation for this integral, but it may immediately be demonstrated that it has several useful properties closely analogous to those of conventional integrals, which will now be demonstrated, beginning with the translational invariance of the integral.

Proposition 11.1.2 *If f is a function in $G^\infty(\mathbb{R}_S^{0,n})$, and η is an element of $\mathbb{R}_S^{0,n}$, then*

$$\int d^n\xi \, f(\xi + \eta) = \int d^n\xi \, f(\xi). \tag{11.3}$$

Proof Suppose that

$$f(\xi) = f_{1\ldots n}(\xi^1 \ldots \xi^n) + \text{ lower order terms.} \tag{11.4}$$

Then

$$f(\xi + \eta) = f_{1\ldots n}(\xi^1 \ldots \xi^n) + g(\xi, \eta) \tag{11.5}$$

where $g(\xi, \eta)$ has no term of top order in ξ. Hence

$$\int d^n\xi \, f(\xi + \eta) = f_{1\ldots n} = \int d^n\xi \, f(\xi). \tag{11.6}$$

∎

The next proposition contains the integration by parts formula.

Proposition 11.1.3 *Suppose that f and g are in $G^\infty(\mathbb{R}_S^{0,n})$. Then*

$$\int d^n\xi \, f \, \partial_j^O g = -(-1)^{|f|} \int d^n\xi \, \partial_j^O f \, g. \tag{11.7}$$

Proof By the graded Leibniz rule Theorem 4.4.1(d),

$$\partial_j^O(fg) = f \, \partial_j^O g + (-1)^{|f|} \partial_j^O f \, g. \tag{11.8}$$

Also, since the expansion of $\partial_j^O(fg)(\xi)$ in powers of ξ can have no top term,

$$\int d^n\xi \, \partial_j^O(fg) = 0. \tag{11.9}$$

∎

It will next be shown that a Fourier transform can be defined which leads to a Fourier inversion theorem.

Definition 11.1.4   If f is in $G^\infty(\mathbb{R}^{0,n}_S)$, then the *Fourier transform* of f is the function $\hat{f}$ of $\mathbb{R}^{0,n}_S$ into $\mathbb{R}_S$ defined by

$$\hat{f}(\rho) = i^{n^2/2} \int d^n\rho \, \exp(-i\rho \cdot \xi) f(\xi) \tag{11.10}$$

where $\rho \cdot \xi = \sum_{j=1}^{n} \rho^j \xi^j$.

The content of the following theorem is that the Fourier transform defines an involution on $G^\infty(\mathbb{R}^{0,n}_S)$.

Theorem 11.1.5

(a) *The Fourier transform $\hat{f}$ of a G^∞ function f is itself G^∞.*

(b) *If $\hat{\hat{f}}$ denotes the Fourier transform of the Fourier transform of f, then*

$$\hat{\hat{f}} = f. \tag{11.11}$$

Proof

(a) Since the integrand in the defining equation (11.10) is a power series jointly in ρ and ξ, the integral itself will be a power series in ρ, so that $\hat{f}$ must be G^∞.

(b) Explicitly calculating,

$$\hat{\hat{f}}(\eta) = i^{n^2} \int d^n\rho \, d^n\xi \, \exp(-i\eta \cdot \rho)\exp(-i\rho \cdot \xi)f(\xi)$$

$$= i^{n^2} \int d^n\xi \, (-1)^{n(n-1)/2} i^n \prod_{j=1}^{n}(\eta^j - \xi^j)f(\xi). \tag{11.12}$$

Using Proposition 11.1.2, it can be seen that

$$\hat{\hat{f}}(\eta) = i^{(n^2+n)}(-1)^{\left(\frac{n^2+n}{2}\right)} \int d^n\xi \prod_{j=1}^{n} \xi^j f(\xi + \eta)$$

$$= f(\eta) \tag{11.13}$$

as required.   ∎

The Fourier inversion theorem can be used to give an explicit construction of the kernel of a differential operator on $G^\infty(\mathbb{R}^{0,n}_S)$, with kernel defined in the following way.

Definition 11.1.6 Suppose that K is a differential operator on $G^\infty(\mathbb{R}_S^{0,n})$. Then the *integral kernel* of K is the unique G^∞ function $\operatorname{Ker} K(\eta, \xi)$ on $\mathbb{R}_S^{0,2n}$ such that

$$Kf(\xi) = \int d^n\eta \operatorname{Ker} K(\eta, \xi) \, f(\eta). \tag{11.14}$$

A differential operator on $G^\infty(\mathbb{R}_S^{0,n})$ may be expanded as

$$K = \sum_{\underline{\mu} \in M_n} \sum_{\underline{\nu} \in M_n} K_{\underline{\mu}\underline{\nu}} \xi^{\underline{\mu}} \partial_{\underline{\nu}}^O. \tag{11.15}$$

This expansion enables one to construct the integral kernel of K.

Proposition 11.1.7 *The integral kernel of the operator K defined in (11.15) is the function $\operatorname{Ker} K$ on $\mathbb{R}_S^{0,2n}$ with*

$$\operatorname{Ker} K(\eta, \xi) = i^{n^2} \int d^n\rho \sum_{\underline{\mu} \in M_n} \sum_{\underline{\nu} \in M_n} i^{\ell(\underline{\nu})} K_{\underline{\mu}\underline{\nu}} \xi^{\underline{\mu}} \rho^{\underline{\nu}} \exp(-i\rho \cdot (\eta - \xi)),$$

$$\tag{11.16}$$

where $\ell(\underline{\nu})$ denotes the number of terms in the multi-index $\underline{\nu}$.

Proof Since, by Theorem 11.1.5,

$$f(\xi) = i^{n^2} \int d^n\rho \, d^n\eta \exp(-i\rho \cdot (\eta - \xi)) f(\eta),$$

$$Kf(\xi) = i^{n^2} \int d^n\rho \, d^n\eta \sum_{\underline{\mu} \in M_n} \sum_{\underline{\nu} \in M_n} i^{\ell(\underline{\nu})} K_{\underline{\mu}\underline{\nu}} \xi^{\underline{\mu}} \rho^{\underline{\nu}} \exp(-i\rho \cdot (\eta - \xi)) f(\eta).$$

$$\tag{11.17}$$

∎

It is evident from this construction that the kernel of a differential operator is always G^∞. A particular example of an integral kernel is the δ function which is the kernel of the identity operator. Using the preceding proposition, it may be seen that the function

$$\delta(\eta - \xi) = i^{n^2} \int d^n\rho \exp(-i\rho \cdot (\eta - \xi))$$

$$= \prod_{i=1}^{n} (\eta^i - \xi^i) \tag{11.18}$$

satisfies the defining property of the delta function, that is,

$$\int d^n \eta \, \delta(\eta - \xi) f(\eta) = f(\xi).$$ (11.19)

These constructions play an important role in fermionic path integrals, as will be seen in Chapter 15.

A differential operator is a linear operator on $G^\infty(\mathbb{R}_S^{0,n})$ regarded as a $(2^{n-1}, 2^{n-1})$ dimensional free $\mathbb{R}_S$ module. It may readily be shown that integration of the kernel of K at coincident points gives the supertrace, that is

$$\mathrm{Str}\, K = \int d^n \xi \, \mathrm{Ker}\, K(\xi, \xi),$$ (11.20)

while the standard trace may be obtained from the formula

$$\mathrm{Tr}\, K = \int d^n \xi \, \mathrm{Ker}\, K(\xi, -\xi).$$ (11.21)

11.2 Integration on $\mathbb{R}_S^{m,n}$

The first step in building a theory of integration on supermanifolds is to define an integral on open subsets of $\mathbb{R}_S^{m,n}$, together with a rule for transforming the integral under G^∞ change of coordinates. The key definition, essentially due to Berezin, defines an integral over an open set U in $\mathbb{R}_S^{m,n}$ in terms of a Berezin integral over anticommuting variables together with an ordinary integral over the body $U_{[\emptyset]}$ of U. This definition leads to a good transformation rule, involving the super determinant of the matrix of the coordinate derivatives, but one whose validity is restricted to functions with compact support or sufficiently fast decay. In later sections of this chapter the integration theory of this section is applied to compact supermanifolds, and then some alternative (but related) approaches introduced to handle the more general case.

Definition 11.2.1   Let U be open in $\mathbb{R}_S^{m,n}$ and $f : U \to \mathbb{R}_S$ be G^∞. Then the integral of f over U is defined to be

$$\int_U d^m x \, d^n \xi \, f(x^1, \dots, x^m; \xi^1, \dots, \xi^n)$$
$$= \int_{U_{[\emptyset]}} d^m x \left(\int d^n \xi \, f(x^1, \dots, x^m; \xi^1, \dots, \xi^n) \right)$$ (11.22)

where the integration with respect to $x^1, \ldots, x^m$ is evaluated as a standard Riemann integral, that is, each coefficient in the expansion of $f_{1\ldots n}$ in terms of Grassmann generators is evaluated as a Riemann integral, the integral being defined only when each of these integrals separately converges.

A simple example will now be given to illustrate this definition.

Example 11.2.2   Suppose that $V = (0,1) \subset \mathbb{R}$ and that $U = (\epsilon_{1,2})^{-1}(V)$. Also let

$$
\begin{aligned}
f : U &\to \mathbb{R}_S \\
(x; \xi^1, \xi^2) &\mapsto x\xi^1 + 10x\xi^1\xi^2.
\end{aligned}
\tag{11.23}
$$

Then

$$
\int_U \mathrm{d}x\, \mathrm{d}^2\xi\, f(x; \xi^1, \xi^2) = \int_0^1 \mathrm{d}x\, 10x = 5.
\tag{11.24}
$$

An important theorem, showing how the integral behaves under change of variable, will now be presented. The result was originally presented by Leĭtes (in the slightly restricted context of H^∞ supermanifolds).

Theorem 11.2.3   *Suppose that U is open in $\mathbb{R}_S^{m,n}$ and that $H : U \to \mathbb{R}_S^{m,n}$ is a superdiffeomorphism of U onto its image $H(U)$. Also suppose that $f : H(U) \to \mathbb{R}_S$ is G^∞ and that the support of f is compact and contained in $H(U)$. Then, if $H(x; \xi) = (y; \eta)$,*

$$
\int_{H(U)} \mathrm{d}^m y\, \mathrm{d}^n \eta\, f(y; \eta) = \int_U \mathrm{d}^m x\, \mathrm{d}^n \xi\, \mathrm{Ber}(H)(x; \xi)\, (f \circ H)(x; \xi)
\tag{11.25}
$$

where

$$
\mathrm{Ber}(H)(x; \xi) = \mathrm{sdet}
\begin{pmatrix}
\frac{\partial y^i}{\partial x^j}(x; \xi) & \frac{\partial \eta^k}{\partial x^j}(x; \xi) \\
\frac{\partial y^i}{\partial \xi^l}(x; \xi) & \frac{\partial \eta^k}{\partial \xi^l}(x; \xi)
\end{pmatrix}.
\tag{11.26}
$$

Proof   As a consequence of the chain rule, if H_1 and H_2 are superdiffeomorphisms of open subsets of $\mathbb{R}_S^{m,n}$ with the image of H_1 equal to the domain of H_2, then

$$
\mathrm{Ber}(H_1)\, \mathrm{Ber}(H_2) = \mathrm{Ber}(H_2 \circ H_1).
\tag{11.27}
$$

Also, any superdiffeomorphism H of an open subset of $\mathbb{R}_S^{m,n}$ may be decomposed as $H = H_2 \circ H_1$ where H_1 has the form

$$
H_1(x; \xi) = (h_1(x; \xi); \xi),
\tag{11.28}
$$

with h_1 a function taking values in $\mathbb{R}_S^{m,0}$, and H_2 is of the form

$$H_2(x;\xi) = (x;\phi_2(x;\xi)) \tag{11.29}$$

with ϕ_2 a function taking values in $\mathbb{R}_S^{0,n}$. It is thus sufficient to prove the theorem for functions of each of these two types.

Suppose that H is of the form

$$H(x;\xi) = (h(x;\xi);\xi), \tag{11.30}$$

where $h : \mathbb{R}_S^{m,n} \to \mathbb{R}_S^{m,0}$. Then

$$\mathrm{Ber}(H)(x;\xi) = \det\left(\frac{\partial h^i}{\partial x^j}\right)(x;\xi). \tag{11.31}$$

Thus if

$$f(y;\eta) = \sum_{\mu \in M_n} f_\mu(y)\eta^\mu, \tag{11.32}$$

then

$$\int_U \mathrm{d}^P y \, \mathrm{d}^Q \eta \, f(y;\eta) = \int_{(\epsilon_{m,0} \circ h)(U_{[\emptyset]})} \mathrm{d}^P y \, f_{1\ldots n}(y) \tag{11.33}$$

while

$$\int_U \mathrm{d}^m x \, \mathrm{d}^n \xi \, \mathrm{Ber}(H)(x;\xi) \, (f \circ H)(x;\xi)$$

$$= \int_{U_{[\emptyset]}} \mathrm{d}^m x \, \frac{\partial}{\partial \xi^n} \cdots \frac{\partial}{\partial \xi^1} \left(\det\left(\frac{\partial h^i}{\partial x^j}\right)(x;\xi) f(h(x;\xi);\xi) \right)$$

$$= \int_{U_{[\emptyset]}} \mathrm{d}^m x \left(f_{1\ldots n}(h(x;0)) \det\left(\frac{\partial h^i}{\partial x^j}\right)(x;0) \right.$$

$$\left. + \frac{\partial}{\partial \xi^n} \cdots \frac{\partial}{\partial \xi^1} \left(\sum_{\mu \in M_n, |\mu| < n} f_\mu(x)\xi^\mu \det\left(\frac{\partial h^i}{\partial x^j}\right)(x;\xi) \right) \right). \tag{11.34}$$

Now, by standard results for integration on $\mathbb{R}^m$,

$$\int_{U_{[\emptyset]}} \mathrm{d}^m x \, f_{1\ldots n}(h(x;0)) \det\left(\frac{\partial h^i}{\partial x^j}\right)(x;0) = \int_{H(U)_{[\emptyset]}} \mathrm{d}^m y \, f_{1\ldots n}(y). \tag{11.35}$$

Also, it can be shown by explicit calculation that the second term on the right hand side of (11.34) is the integral of a total derivative, and thus (because of the condition on the support of f) this term must be zero. This establishes the result for the case when the function H has the form H_1.

Turning now to the other case, suppose that

$$H(x;\xi) = (x;\phi(x;\xi)).$$ (11.36)

It must be shown that

$$\int_U \mathrm{d}^m x\, \mathrm{d}^n \eta\, f(x;\eta) = \int_U \mathrm{d}^m x\, \mathrm{d}^n \xi\, f(x;\phi(x;\xi)) \times \frac{1}{\det\left(\frac{\partial\phi^i}{\partial\xi^j}\right)}.$$ (11.37)

This may be shown by expressing the function ϕ as a combination of $n+1$ steps, $\phi = \phi_1 \circ \cdots \circ \phi_{n+1}$ where for $r = 1, \ldots, n$, ϕ_r is of the form

$$\phi_r(x;\xi) = (\xi^1, \ldots, \xi^{r-1}, \xi^r k^r(x;\xi^1, \ldots, \hat{\xi}^r, \ldots \xi^n), \xi^{r+1}, \ldots, \xi^n),$$ (11.38)

while ϕ_{n+1} is linear in ξ. It must then be shown that (11.37) holds when ϕ is replaced by any of the ϕ_r, $r = 1, \ldots, n+1$. Now, for $r = 1, \ldots n$,

$$\int_U \mathrm{d}^m x\, \mathrm{d}^n \xi\, f(x;\phi_r(x;\xi)) \times \frac{1}{\det\left(\frac{\partial\phi^i_r}{\partial\xi^j}\right)}$$

$$= \int_U \mathrm{d}^m x\, \mathrm{d}^n \xi\, f(x;\phi_r(x;\xi)) \times \frac{1}{k^r(x;\xi^1, \ldots, \hat{\xi}^r, \ldots \xi^n)}$$

$$= \int_U \mathrm{d}^m x\, \mathrm{d}^n \xi\, f_{1\ldots n}(x)\xi^1 \ldots \xi^n \times \frac{k^r(x;\xi^1, \ldots, \hat{\xi}^r, \ldots \xi^n)}{k^r(x;\xi^1, \ldots, \hat{\xi}^r, \ldots \xi^n)}$$

$$= \int_U \mathrm{d}^m x\, \mathrm{d}^n \eta\, f(x;\eta).$$ (11.39)

Finally it must be shown that (11.37) holds when ϕ is replaced by ϕ_{n+1}. Suppose that

$$\phi_{n+1}(x;\xi) = (a^1{}_j(x)\xi^j + \gamma^1(x), \ldots, a^n{}_j(x)\xi^j + \gamma^n(x)).$$ (11.40)

Then

$$\int_U \mathrm{d}^m x\, \mathrm{d}^n \xi\, f(x;\phi(x;\xi)) \times \frac{1}{\det\left(\frac{\partial\phi^i_{n+1}}{\partial\xi^j}\right)}$$

$$= \int_U \mathrm{d}^m x\, \mathrm{d}^n \xi\, f_{1\ldots n}(x)\xi^1 \ldots \xi^n \frac{\det\left(a^i{}_j(x)\right)}{\det\left(a^i{}_j(x)\right)}$$

$$= \int_U \mathrm{d}^m x\, \mathrm{d}^n \eta\, f(x;\eta).$$ (11.41)

11.3   Integration on compact supermanifolds

The method of integration on $\mathbb{R}_S^{m,n}$ developed in the previous section leads naturally to an integral of a Berezin density on a compact supermanifold. As with conventional manifolds, a partition of unity is used to sum the contribution from different coordinate patches.

Definition 11.3.1   Suppose that ω is a Berezin density on $\mathcal{M}$ and that the collection $\{(V_\alpha, f_\alpha)|\alpha \in \Gamma\}$ is a partition of unity on $\mathcal{M}$ where each V_α is a coordinate neighbourhood with corresponding coordinate map ψ_α. Then the integral of ω over $\mathcal{M}$ is defined to be

$$\int_{\mathcal{M}} \omega = \sum_{\alpha \in \Gamma} \int_{\psi_\alpha(V_\alpha)} \mathrm{d}^m x \, \mathrm{d}^n \xi \, \omega_\alpha(x; \xi) \, f_\alpha \circ \psi_\alpha^{-1}(x; \xi), \qquad (11.42)$$

where (as in Definition 10.2.1) ω_α is the local representative of ω in the chart (V_α, ψ_α).

It is of course essential to show that the value of this integral is independent of the choice of partition of unity chosen. This result is the content of the following theorem, which combines the transformation rule for a Berezinian density with the transformation rule for integrals (Theorem 11.2.3).

Theorem 11.3.2   *Suppose that $\{(V_\alpha, f_\alpha)|\alpha \in \Gamma\}$ and $\{(U_\beta, g_\beta)|\beta \in \Xi\}$ are partitions of unity on $\mathcal{M}$, with each $V_\alpha, \alpha \in \Gamma$ and each $U_\beta, \beta \in \Xi$ a coordinate neighbourhood (with corresponding coordinate maps ψ_α and ϕ_β). Then*

$$\sum_{\alpha \in \Gamma} \int_{\psi_\alpha(V_\alpha)} \mathrm{d}^m x \, \mathrm{d}^n \xi \, \omega_\alpha(x; \xi) \, f_\alpha \circ \psi_\alpha^{-1}(x; \xi)$$

$$= \sum_{\beta \in \Xi} \int_{\phi_\beta(U_\beta)} \mathrm{d}^m x \, \mathrm{d}^n \xi \, \omega_\beta(y; \eta) \, g_\beta \circ \phi_\beta^{-1}(y; \eta). \qquad (11.43)$$

Proof   Let $W_{\alpha\beta} = V_\alpha \cap U_\beta$. Then

$$\sum_{\alpha \in \Gamma} \int_{\psi_\alpha(V_\alpha)} \mathrm{d}^m x \, \mathrm{d}^n \xi \, \omega_\alpha(x; \xi) f_\alpha \circ \psi_\alpha^{-1}(x; \xi)$$

$$= \sum_{\alpha \in \Gamma, \beta \in \Xi} \int_{\psi_\alpha(W_{\alpha\beta})} \mathrm{d}^m x \, \mathrm{d}^n \xi \omega_\alpha(x; \xi) f_\alpha \circ \psi_\alpha^{-1}(x; \xi) g_\beta \circ \psi_\alpha^{-1}(x; \xi)$$

$$= \sum_{\alpha \in \Gamma, \beta \in \Xi} \int_{\phi_\beta(W_{\alpha\beta})} \mathrm{d}^m y \, \mathrm{d}^n \eta \, \omega_\beta(y; \eta) f_\alpha \circ \phi_\beta^{-1}(y; \eta) g_\beta \circ \phi_\beta^{-1}(y; \eta),$$

$$(11.44)$$

using Theorem 11.2.3 and the fact that

$$\omega_\alpha = \omega_\beta \times \mathrm{Ber}(\psi_\alpha \circ \phi_\beta^{-1}). \qquad (11.45)$$

Also

$$\sum_{\beta \in \Xi} \int_{\psi_\beta(U_\beta)} \mathrm{d}^m y\, \mathrm{d}^n \eta\, \omega_\beta(x;\xi) g_\beta \circ \phi_\beta^{-1}(x;\xi)$$

$$= \sum_{\alpha \in \Gamma, \beta \in \Xi} \int_{\phi_\beta(W_{\alpha\beta})} \mathrm{d}^m y\, \mathrm{d}^n \eta\, \omega_\beta(x;\xi) f_\alpha \circ \phi_\beta^{-1}(y;\eta) g_\beta \circ \phi_\beta^{-1}(y;\eta). $$
$$(11.46)$$

∎

This theorem cannot be extended to noncompact supermanifolds, or to finite regions, except when considering functions which die away sufficiently fast on the boundary of the region of integration. A simple example of the breakdown of Theorem 11.2.3 illustrates the problem:

Example 11.3.3 Let f be the function in $G^\infty(\mathbb{R}_S^{1,2})$ with

$$f(y; \eta^1, \eta^2) = y. \qquad (11.47)$$

Then, if I denotes the unit interval $[1, 0]$ of the real line,

$$\int_I \mathrm{d}y\, \mathrm{d}^2 \eta\, f(y; \eta) = 0. \qquad (11.48)$$

Changing coordinates on $\mathbb{R}_S^{1,2}$ by the diffeomorphism

$$H(x;\xi) = (y(x;\xi); \eta(x;\xi)) \qquad (11.49)$$

with

$$y(x;\xi) = x + \xi^1 \xi^2$$
$$\eta^i(x;\xi) = \xi^i \qquad i = 1, 2 \qquad (11.50)$$

the rule in theorem Theorem 11.2.3 suggests that one should find that

$$\int_I \mathrm{d}y\, \mathrm{d}^2 \eta\, f(y;\eta) = \int_I \mathrm{d}x\, \mathrm{d}^2 \xi\, f(H(x;\xi)) \qquad (11.51)$$

since the Berezinian of H is equal to 1. However in fact

$$\int_I \mathrm{d}x\, \mathrm{d}^2 \xi\, f(H(x;\xi)) = \int_I \mathrm{d}x\, \mathrm{d}^2 \xi\, (x + \xi^1 \xi^2)$$
$$= 1 \qquad (11.52)$$

and so the rule of Theorem 11.2.3 breaks down in this case.

From this example it is clear that simply treating the region of integration as part of $\mathbb{R}^m$ can lead to inconsistencies. A cure may be found by considering the even part of an integral on $\mathbb{R}^{m,n}_S$ to be a contour integral in $\mathbb{R}^{m,0}_S$, but the combination of this with the Berezin integral is rather ad-hoc [43, 121]. In the following sections further approaches to integration on supermanifolds, taking the theory beyond the compact situation, are considered. In Rothstein's approach, described in the next section, the Berezin integral is treated as a differential operator, while Voronov's approach in Section 11.5 considers regions in $\mathbb{R}^{m,n}_S$ rather than simply in $\mathbb{R}^m$.

11.4   Rothstein's theory of integration on non-compact supermanifolds

It is evident from Example 11.3.3 that the theory of integration may break down for non-compact regions of integration, or for regions with boundaries, because of the breakdown of the transformation rule of Theorem 11.2.3. In this section an elegant approach to integration on non-compact supermanifolds developed by Rothstein is described. The key idea in this approach is a new characterization of the Berezinian in terms of differential-operator valued forms.

Throughout this section $\mathcal{M}$ is a $(\mathbb{R}^{m,n}_S, \mathrm{DeWitt}, G^\infty)$ supermanifold with orientable body. Suppose that U is an open subset of $\mathcal{M}$. A differential operator on U is an operator L on $G^\infty(U)$ which can be expressed locally as a finite sum of the form

$$L = \sum_{\underline{I} \in N_m} \sum_{\underline{\mu} \in M_n} f_{\underline{I}\,\underline{\mu}}(x;\xi) \frac{\partial}{\partial x^{\underline{I}}} \frac{\partial}{\partial \xi^{\underline{\mu}}} \tag{11.53}$$

where N_m is the set of multi-indices $\underline{I} = I_1 \ldots I_k$ with $1 \leq I_i \leq m, i = 1, \ldots k$, together with the empty multi-index $\emptyset$, while

$$\frac{\partial}{\partial x^{\underline{I}}} = \frac{\partial}{\partial x^{I_1}} \cdots \frac{\partial}{\partial x^{I_k}}, \tag{11.54}$$

and $\frac{\partial}{\partial x^{\emptyset}}$ is the identity operator. The space $\mathfrak{O}(U)$ of differential operators on U has a natural left graded $G^\infty(U)$-module structure with

$$(fL)(g) = f\,L(g) \tag{11.55}$$

where f and g are in $G^\infty(U)$. The differential operator L may also be expanded as

$$L = \sum_{\underline{I} \in N_m} \sum_{\underline{\mu} \in M_n} \frac{\partial}{\partial x^{\underline{I}}} \frac{\partial}{\partial \xi^{\underline{\mu}}} \circ g_{\underline{I}\,\underline{\mu}}(x; \xi) \qquad (11.56)$$

with $\circ$ denoting combination of operators. The notion of Rothstein form on the (m, n)-dimensional supermanifold $\mathcal{M}$ will now be defined.

Definition 11.4.1 Let U be an open subset of $\mathcal{M}$.

(a) The set $\Omega^m(U_{[\emptyset]})$ of m-forms on the body of U is given a left $G^\infty(U)$-module structure by defining

$$f\omega = f_{[\emptyset]}\omega \qquad (11.57)$$

for each f in $G^\infty(U)$ and each ω in $\Omega^m(U_{[\emptyset]})$.

(b) A *Rothstein form* on U is an element of $\Omega^m(U_{[\emptyset]}) \otimes_{G^\infty(U)} \mathfrak{D}(U)$.

(c) The set of Rothstein forms on U is denoted $\mathfrak{R}(U)$.

Elements of $\mathfrak{R}(U)$ act on functions in $G^\infty(U)$ to give m-forms on $U_{[\emptyset]}$. This action will be denoted by square brackets, so that if $R = \omega \otimes L$ is in $\mathfrak{R}(U)$ and f is in $G^\infty(U)$, then $R[f]$ is the m-form on $U_{[\emptyset]}$ defined by

$$R[f] = L(f)\omega. \qquad (11.58)$$

This action involves the left product of Definition 11.4.1(a), so that any ξ-dependent terms in $L(f)$ are set to zero.

The space $\mathfrak{R}(U)$ has the structure of both a left and right $G^\infty(U)$-module. The left $G^\infty(U)$-module structure is defined by

$$(fR)[g] = f\,R[g], \qquad (11.59)$$

again using the left product of Definition 11.4.1(a). The right module structure is given by

$$(Rf)[g] = R[fg] \qquad (11.60)$$

and is the structure which will be used in the following.

It follows from the definition that if V is a coordinate neighbourhood of $\mathcal{M}$ then the Rothstein form R can be expressed as a sum of the form

$$R = \sum_{\underline{I} \in N_m} \sum_{\underline{\mu} \in M_n} dx^1 \wedge \cdots \wedge dx^m \otimes \frac{\partial}{\partial x^{\underline{I}}} \frac{\partial}{\partial \xi^{\underline{\mu}}} \circ f_{\underline{I}\,\underline{\mu}}(x; \xi). \qquad (11.61)$$

However the set A with

$$A = \left\{ dx^1 \wedge \cdots \wedge dx^m \otimes \frac{\partial}{\partial x^{\underline{I}}} \frac{\partial}{\partial \xi^{\underline{\mu}}} | \underline{I} \in N_m, \underline{\mu} \in M_n \right\}$$

does not form a basis for the right $G^\infty(V)$-module $\mathfrak{R}(V)$, as may be seen (for example) by observing that

$$dx^1 \wedge \cdots \wedge dx^m \otimes \frac{\partial}{\partial \theta^1} \circ \theta^1 \theta^2 = 0. \tag{11.62}$$

None the less, as is established in the following theorem, $\mathfrak{R}(V)$ is in fact a free module if V is a coordinate neighbourhood; its rank is infinite, and one possible basis is a subset of the set A.

Theorem 11.4.2   *The space $\mathfrak{R}(V)$ of Rothstein forms on a coordinate neighbourhood V in $\mathcal{M}$ is a free right $G^\infty(V)$-module. The set B where*

$$B = \left\{ dx^1 \wedge \cdots \wedge dx^m \otimes \frac{\partial}{\partial x^{\underline{I}}} \frac{\partial^n}{\partial \xi^n \ldots \partial \xi^1} | \underline{I} \in N_m \right\} \tag{11.63}$$

is a basis for this module.

Proof   An arbitrary element R of $\mathfrak{R}(V)$ can be expressed as

$$R = \sum_{\underline{I} \in N_m} \sum_{\underline{\mu} \in M_n} dx^1 \wedge \cdots \wedge dx^m \otimes \frac{\partial}{\partial x^{\underline{I}}} \frac{\partial}{\partial \xi^{\underline{\mu}}} \circ f_{\underline{I}\underline{\mu}}(x; \xi). \tag{11.64}$$

However, because left multiplication of forms on $\Omega^m(U_{[\emptyset]})$ by functions in $G^\infty(V)$ annihilates the anticommuting coordinates,

$$dx^1 \wedge \cdots \wedge dx^m \otimes \frac{\partial}{\partial \xi^{\underline{\mu}}} \circ \xi^{\underline{\rho}} = 0 \tag{11.65}$$

unless ρ is a sub-multi-index of μ. Also

$$dx^1 \wedge \cdots \wedge dx^m \otimes \frac{\partial^n}{\partial \xi^n \ldots \partial \xi^1} \circ \xi^{\underline{\mu}} = \pm dx^1 \wedge \cdots \wedge dx^m \otimes \frac{\partial}{\partial \xi^{\underline{\mu}_c}} \tag{11.66}$$

where $\underline{\mu}_c$ is the complementary multi-index to $\underline{\mu}$ in the sense that $\xi^{\underline{\mu}} \xi^{\underline{\mu}_c} = \pm \xi^1 \ldots \xi^n$. Thus B spans $\mathfrak{R}(V)$.

   That B is linearly independent follows from the linear independence of $\left\{ \frac{\partial}{\partial x^{\underline{I}}} | \underline{I} \in N_m \right\}$ together with the fact that

$$dx^1 \wedge \cdots \wedge dx^m \frac{\partial}{\partial x^{\underline{I}}} \frac{\partial^n}{\partial \xi^n \ldots \partial \xi^1} \circ f(x) \xi^{\underline{\mu}} [\xi^{\underline{\mu}_c}] = \pm dx^1 \wedge \cdots \wedge dx^m \frac{\partial}{\partial x^{\underline{I}}} f(x). \tag{11.67}$$

Hence B is a basis of $\mathfrak{R}(V)$, which is thus a free module. ∎

The next step is to define the integral of a Rothstein form over $\mathcal{M}$, and then to define a quotient space of $\mathfrak{R}(U)$ which corresponds to the space $\mathfrak{B}er(U)$ of Berezinian forms on U, so that the usual Berezin integral is recovered.

Definition 11.4.3   Let R be an element of $\mathfrak{R}(\mathcal{M})$. Then the integral of R over $\mathcal{M}$ is defined to be

$$\int_{\mathcal{M}} R = \int_{\mathcal{M}_{[\emptyset]}} R[1]. \tag{11.68}$$

A simple example of this integral, which demonstrates its relation to the Berezin integral, is given by the integral of $R(x;\xi) = \mathrm{d}x^1 \wedge \cdots \wedge \mathrm{d}x^m \otimes 1\frac{\partial^n}{\partial\xi^n \ldots \partial\xi^1} \circ f(x;\xi)$ over an open subset V of $\mathbb{R}_S^{m,n}$; this takes the form

$$\int_V R = \int_{V_{[\emptyset]}} \mathrm{d}x^1 \wedge \cdots \wedge \mathrm{d}x^m \frac{\partial^n f(x;\xi)}{\partial\xi^n \ldots \partial\xi^1}. \tag{11.69}$$

The following theorem establishes the quotient space of $\mathfrak{R}(U)$ which corresponds to $\mathfrak{B}er(U)$ and leads to the Berezin integral.

Theorem 11.4.4   *Suppose that, for each open subset U of $\mathcal{M}$, $\mathfrak{R}_+(U)$ is defined to be the sub-module of of $\mathfrak{R}(U)$ consisting of elements R such that*

$$\int_{U_{[\emptyset]}} R[h] = 0 \tag{11.70}$$

whenever h is a compact support function in $G^\infty(U)$. Then, if V is a co-ordinate neighbourhood of $\mathcal{M}$,

(a) R is in $\mathfrak{R}_+(V)$ if an only if

$$R = \sum_{\underline{I} \in N_m} \mathrm{d}x^1 \wedge \cdots \wedge \mathrm{d}x^m \frac{\partial}{\partial x^{\underline{I}}} \frac{\partial^n}{\partial\xi^n \ldots \partial\xi^1} f^{\underline{I}}(x;\xi) \tag{11.71}$$

   with $f^\emptyset = 0$.
(b) If R is in $\mathfrak{R}_+(V)$, then $R[1]$ is exact.
(c) The quotient module $\mathfrak{R}(V)/\mathfrak{R}_+(V)$ is locally free with dimension 1 and basis
   $$\left\{ \mathrm{d}x^1 \wedge \cdots \wedge \mathrm{d}x^m 1\frac{\partial^n}{\partial\xi^n \ldots \partial\xi^1} \right\}.$$

Proof (a) Suppose that R is in $\mathfrak{R}(V)$, with

$$R = \sum_{\underline{I} \in N_m} \mathrm{d}x^1 \wedge \cdots \wedge \mathrm{d}x^m \frac{\partial}{\partial x^{\underline{I}}} \frac{\partial^n}{\partial \xi^n \ldots \partial \xi^1} f^{\underline{I}}(x; \xi)$$

and that $f^{\emptyset} \neq 0$. Then there must exist a compact support function h in $G^{\infty}(V)$ such that

$$\int_{V_{[\emptyset]}} \mathrm{d}x^1 \wedge \cdots \wedge \mathrm{d}x^m \frac{\partial^n}{\partial \xi^n \ldots \partial \xi^1} f^{\emptyset} h \neq 0. \qquad (11.72)$$

Thus

$$\int_{V_{[\emptyset]}} R[h\xi^1 \ldots \xi^n] \neq 0 \qquad (11.73)$$

so that R cannot be in $\mathfrak{R}_+$. Conversely, if $f^{\emptyset} = 0$, and h is any compact support function on $G^{\infty}(V)$, $\int_{V_{[\emptyset]}} R[h]$ consists of terms of the form

$$\int_{V_{[\emptyset]}} \mathrm{d}x^1 \wedge \cdots \wedge \mathrm{d}x^m \frac{\partial f_{\underline{I}} h_{1 \ldots n}}{\partial x^{\underline{I}}} (x)$$

with $\underline{I} \neq 0$, all of which are zero.

(b) If R is in $\mathfrak{R}_+(V)$, then $R[1]$ consists of terms of the form

$$\int_{V_{[\emptyset]}} \mathrm{d}x^1 \wedge \cdots \wedge \mathrm{d}x^m \frac{\partial f_{\underline{I}}}{\partial x^{\underline{I}}} (x)$$

with $\underline{I} \neq 0$, each of which is exact. Part (c) is an immediate consequence of part (a).

Corollary 11.4.5 *The sheaves $\mathfrak{R}/\mathfrak{R}_+$ and $\mathfrak{Ber}$ over $\mathcal{M}_{[\emptyset]}$ are isomorphic.*

An alternative characterisation of $\mathfrak{Ber}$ which is closely related to Rothstein's construction may be found in the work of Penkov [108]. In Rothstein's approach to integration on supermanifolds, the integral of an element R of $\mathfrak{R}(\mathcal{M})$ is defined to be

$$\int_{\mathcal{M}} R = \int_{\mathcal{M}_{[\emptyset]}} R[1]. \qquad (11.74)$$

For Rothstein forms of compact support, this definition is precisely equivalent to the definition of integration given in the preceding section, since a well-defined definition of integration on $\mathfrak{R}/\mathfrak{R}_+$ is recovered. For more general Rothstein forms, this is not the case, but it is now possible to calculate the necessary extra terms to go in to the transformation law, bearing in mind that the Rothstein definition of integration is manifestly intrinsically defined. If $\omega_1(x; \xi)$ is the local representative of a Berezinian form ω in local coordinates $(x; \xi)$, then this is related to its local representative $\omega_2(y; \eta)$ in coordinates $(y; \eta)$ by the rule

$$\omega_2(y; \eta) = \omega_1(x; \xi) \mathrm{Ber} \left(\frac{y, \eta}{x, \xi} \right) \tag{11.75}$$

while the Rothstein form R with local expansion

$$\mathrm{d}y^1 \wedge \cdots \wedge \mathrm{d}y^m 1 \frac{\partial^n}{\partial \eta^n \dots \partial \eta^1} \omega_2(y; \eta)$$

in the coordinates $(y; \eta)$ has an expansion of the form

$$R = \mathrm{d}x^1 \wedge \cdots \wedge \mathrm{d}x^m \frac{\partial^n}{\partial \xi^n \dots \partial \xi^1} \omega(x; \xi)$$

$$+ \sum_{\underline{I} \in N_m/0} \mathrm{d}x^1 \wedge \cdots \wedge \mathrm{d}x^m \frac{\partial}{\partial x^{\underline{I}}} \frac{\partial^n}{\partial \xi^n \dots \partial \xi^1} f^{\underline{I}}(x; \xi)$$

$$\tag{11.76}$$

in the $(x; \xi)$ system, and the extra terms may contribute to the integral in the transformed coordinates.

As an example it will be shown that the transformation rule for Rothstein forms solves the problem of Example 11.3.3. Using Rothstein forms, where previously $\mathrm{d}y\, \mathrm{d}^2\eta$ was transformed to $\mathrm{d}x\, \mathrm{d}^2\xi$, it can be seen that $\mathrm{d}y 1 \frac{\partial}{\partial \eta^1} \frac{\partial}{\partial \eta^2}$ is transformed to

$$\mathrm{d}x \left(1 \frac{\partial}{\partial \xi^2} + \xi^1 1 \frac{\partial}{\partial x} \right) \left(1 \frac{\partial}{\partial \xi^1} - \xi^2 1 \frac{\partial}{\partial x} \right),$$

so that the transformed integral becomes

$$\int_I \mathrm{d}x \left(\frac{\partial}{\partial \xi^2} + \xi^1 \frac{\partial}{\partial x} \right) \left(\frac{\partial}{\partial \xi^1} - \xi^2 \frac{\partial}{\partial x} \right) (x + \xi^1 \xi^2) = \int_I \mathrm{d}x (1 - 1) = 0 \tag{11.77}$$

as required. A more complete and systematic development of these ideas may be found in Rothstein's paper [133].

11.5 Voronov's theory of integration of super forms

Recall from Section 10.4 that an (r, s) super form $\mathcal{L}$ on an (m, n) dimensional supermanifold locally has the form $\mathcal{L}_\alpha(X^i, \dot{X}^i_{a_\alpha})$ with

$$\mathcal{L}_\beta(X^i, \dot{X}^i_{a_\beta}) = \text{Ber}\left(\psi_\beta \circ \psi_\alpha^{-1}\right) \mathcal{L}_\alpha(X^i, \dot{X}^i_{a_\alpha}). \tag{11.78}$$

This object involves a mapping $X : I^{r,s} \to \mathcal{M}$ which can be regarded as a supersmooth (r, s)-cube in $\mathcal{M}$. Its construction has been motivated by the following definition.

Definition 11.5.1

(a) A supersmooth (r, s)-cube on a supermanifold $\mathcal{M}$ is a G^∞ map $X :$ $I^{r,s} \to \mathcal{M}$

(b) The integral of the (r, s)-super form $\mathcal{L}$ over the (r, s)-cube X is

$$\int_X \mathcal{L} = \int \mathrm{d}^r t \mathrm{d}^s \tau \, \mathcal{L}_\alpha(X^i(t; \tau), \dot{X}^i_{a_\alpha}(t; \tau)). \tag{11.79}$$

For compact supermanifolds this leads to a definition of integration essentially equivalent to that of Section 11.3. Two interesting features of Voronov's approach are that it can be extended to include regions with boundary, and that it allows some progress towards an analogue of de Rham theory, because of a Stokes' theorem relating to the exterior derivative (10.28). Voronov's approach to integration over a region with boundary, which takes some inspiration from Berezin [20], defines a region A in $\mathbb{R}^{m,n}_S$ by a condition expressed as

$$u(x; \xi) < 0 \tag{11.80}$$

where u is an even G^∞ function on $\mathbb{R}^{m,n}_S$. The meaning of this condition, which is not straightforward given that u takes values in $\mathbb{R}_S$, is explained below. The boundary is then defined as the set where $u(x; \xi) = 0$. This allows the definition of the integral of a function over A to be given as

$$\int_A \mathrm{d}^m x \mathrm{d}^n \xi \, f(x; \xi) = \int_{\mathbb{R}^{m,n}_S} \mathrm{d}^m x \mathrm{d}^n \xi \, f(x; \xi) \left(-\widehat{\theta}(u(x; \xi))\right) \tag{11.81}$$

where θ is the Grassmann analytic continuation of the Heaviside function on $\mathbb{R}$ which satisfies $\theta(t) = 0$ when $t < 0$ and $\theta(t) = 1$ when $t > 0$, and the Grassmann analytic continuation is constructed using a smooth approximation to θ. It is in this sense that the condition $u(x; \xi) < 0$ is applied.

An example given by Voronov shows how an integral similar to the problem integral of Example 11.3.3 is handled in this approach. This example related to the integration of a function f over the interval $I = [0,1]$ of the real line before and after carrying out the transformation $H(x;\xi) = (y(x;\xi);\eta(x;\xi))$. In Voronov's example the region of integration for the first version of the integral would be specified by the condition $y < 0$ and the integral evaluated is

$$\int_{y<0} dy\, d^2\eta \ \left(f_{[\emptyset]}(y) + f_1(y)\eta^1 + f_2(y)\eta^2 + f_{12}(y)\eta^1\eta^2\right)$$

$$= \int_{\mathbb{R}_S^{1,2}} dy\, d^2\eta \ \left(f_{[\emptyset]}(y) + f_1(y)\eta^1 + f_2(y)\eta^2 + f_{12}(y)\eta^1\eta^2\right)\theta(-y)$$

$$= \int_\infty^0 dy\, f_{12}(y)\,. \tag{11.82}$$

After the change of coordinates corresponding to the diffeomorphism

$$H(x;\xi) = (y(x;\xi);\eta(x;\xi)) \tag{11.83}$$

with

$$y(x;\xi) = x + \xi^1\xi^2$$

$$\eta^i(x;\xi) = \xi^i \qquad i = 1,2 \tag{11.84}$$

the defining condition becomes $x+\xi^1\xi^2 < 0$ so that the integral with respect to the new variables has the form

$$\int_{x+\xi^1\xi^2<0} dx\, d^2\xi \ \left(f_{[\emptyset]}(x + \xi^1\xi^2) + f_1(x)\eta^1 + f_2(x)\eta^2 + f_{12}(x)\eta^1\eta^2\right)$$

$$= \int_{\mathbb{R}_S^{1,2}} dx\, d^2\xi \ \left(f_{[\emptyset]}(x + \xi^1\xi^2) + f_1(x)\xi^1 + f_2(x)\xi^2 + f_{12}(x)\xi^1\xi^2\right)$$

$$\times\, \theta(-(x + \xi^1\xi^2))$$

$$= \int_{\mathbb{R}_S^{1,2}} dx\, d^2\xi \ \left(f_{[\emptyset]}(x) + \xi^1\xi^2\partial_1^E f_{[\emptyset]}(x) + f_1(x)\xi^1 + f_2(x)\xi^2 + f_{12}(x)\xi^1\xi^2\right)$$

$$\times\, \left(\theta(-x) - \xi^1\xi^2\delta(x)\right)$$

$$= \int_\infty^0 \left(f_{12}(x) + \partial_1^E f_{[\emptyset]}(x)\theta(-x) - f_{[\emptyset]}(x)\delta(x)\right)$$

$$= \int_\infty^0 f_{12}(x)\,. \tag{11.85}$$

This illustrates this method of specifying a region of integration, and how it leads to an invariant definition. It is possible to use Voronov's superforms to provide an approach to integration over (r, s)-dimensional regions in a supermanifold. However the theory has not yet been developed to include a full homological algebra of super chains dual to superforms, although the work of Voronov [153] suggests how this development might be made. Voronov's work includes a proof of Stoke's theorem for the exterior derivative (10.28) and boundary of the region defined by $u(x; \xi) < 0$ defined by the condition $u(x; \xi) = 0$. More recently Voronov has introduced the notion of dual form on a supermanifold [154], involving copaths, with the aim of making possible the study of homotopy problems of stable forms and the de Rham cohomology of a supermanifold.

11.6 Integration on $(1, 1)$-dimensional supermanifolds

In this section a somewhat special notion of integration is described, which possesses some of the features of classical integration which are absent in much super integration. The integral is defined with both odd and even limits, and there is an analogue of the fundamental theorem of calculus involving these limits. The integral developed here is used to define contour integration on super Riemann surfaces in Section 14.4.

Definition 11.6.1   Suppose that $g : \mathbb{R}_S^{1,1} \to \mathbb{R}_S$, that $T = (t; \tau)$ are local coordinates on $\mathbb{R}_S^{1,1}$ and that $(a; \alpha)$ and $(b; \beta)$ are points in $\mathbb{R}_S^{1,1}$. Then the integral of g between these two points is defined to be

$$\int_\alpha^\beta \int_a^b g(t, \tau)\mathcal{D}T = \int \left[\int_{a+\tau\alpha}^{b+\tau\beta} g(t, \tau)\mathrm{d}t \right] \mathrm{d}\tau . \qquad (11.86)$$

An important property of this integral is that it obeys what might be called the square root of the fundamental theorem of calculus, that is, if $\mathcal{D}_T$ denotes the super derivative $\mathcal{D}_T = \frac{\partial}{\partial \tau} + \tau \frac{\partial}{\partial t}$ (which satisfies $\mathcal{D}_T^2 = \frac{\partial}{\partial t}$) then

$$\int_\alpha^\beta \int_a^b \mathcal{D}_T g(t, \tau)\mathcal{D}T = g(b; \beta) - g(a; \alpha) . \qquad (11.87)$$

Additionally this integral has the following consistent transformation rule: suppose that the function $F : \mathbb{R}_S^{1,1} \to \mathbb{R}_S^{1,1}, (t; \tau) \mapsto (s(t; \tau); \sigma(t; \tau))$ is real superconformal, that is, there exist smooth functions $f : \mathbb{R} \to \mathbb{R}_{S1}$ and

$\phi : \mathbb{R} \to \mathbb{R}_{S0}$ such that

$$F(t;\tau) = (F_0(t;\tau); F_1(t;\tau))$$
$$= \left(f(t) + \tau\phi(t)\sqrt{f'(t)}, \phi(t) + \sqrt{f'(t) + \phi(t)\phi'(t)} \right) . \ (11.88)$$

(The motivation for this definition can be found in Chapter 14.) Then, if $(c;\gamma) = F(a;\alpha)$ and $(d;\delta) = F(b;\beta)$,

$$\int_\gamma^\delta \int_c^d g(s,\sigma)\mathcal{D}S \int_\alpha^\beta \int_a^b g(F(t;\tau))\mathcal{D}_T F_1(t;\tau)\mathcal{D}T . \qquad (11.89)$$

This result follows from Definition 11.6.1 and the chain rule for the super derivative of combinations of superconformal functions which takes the form

$$\mathcal{D}_T \left(F\left(K(t;\tau) \right) \right) = \mathcal{D}_K F(k(t;\tau); \kappa(t;\tau)) \mathcal{D}_T \left(\kappa(t;\tau) \right) . \qquad (11.90)$$

In Chapter 14 this integral will be used to develop a theory of contour integration on super Riemann surfaces.

11.7 Integration of exterior forms

In this brief final section the integral of an exterior p-form on a supermanifold is defined in terms of maps from $\mathbb{R}^p$ very much as on a classical manifold. At first site this might seem a pointless, if correct, development. However in Chapter 13 this construction will be used to define Lagrangians for supersymmetric extended objects, using an ingenious technique of Howe, Raetzel and Sezgin [77]. The full G^∞ setting is required to give any results distinct from those obtainable from the body of $\mathcal{M}$.

Definition 11.7.1 Suppose that α is an exterior p-form on a supermanifold $\mathcal{M}$ and that $c : I^p \to \mathcal{M}$ is smooth. (As before, I^p denotes the unit p-cube in $\mathbb{R}^p$.) Then $c^*\alpha$ is an $\mathbb{R}_S$-valued p-form on $\mathcal{M}$ and the integral of α over c is defined to be

$$\int_c \alpha = \int_{I^p} c^*\alpha . \qquad (11.91)$$

This notion of integration can immediately be extended to give a theory of integration over smooth p-chains in $\mathcal{M}$, with the usual properties such as Stokes' theorem.

The simplest examples of such integrals use chains whose constituent p-cubes in split local coordinates take the form

$$c(t^1, \ldots, t^p) = (c^1(t^1, \ldots, t^p), \ldots, c^m(t^1, \ldots, t^p); 0, \ldots, 0) \qquad (11.92)$$

with $\epsilon(c^i(t^1, \ldots, t^p)) = c^i(t^1, \ldots, t^p), i = 1, \ldots, m$ so that one is effectively integrating over a submanifold of the body of $\mathcal{M}$, or, with appropriately constructed m-cubes, over the body itself.

Chapter 12

Geometric structures on supermanifolds

In this chapter various geometric structures which can be put on supermanifolds are described. Using the concrete approach and the machinery set up in earlier chapters, much of the material in this chapter is a straightforward analogue of the classical setup.

The chapter begins with the concepts of super principal G bundle (where G is a super Lie group) and associated super bundle, together with the notion of connection and curvature. In section Section 12.2 the particular case is considered of the super bundle of frames of a supermanifold and its associated bundles and the corresponding tensors and densities. In further sections Riemannian and even symplectic structures are considered, while in the final section 12.5 odd symplectic structures, which have no classical analogue, are considered.

12.1 Fibre bundles

In this section the concept of super principal bundle and associated super bundle are defined. The starting point is the definition of super principal bundle.

Definition 12.1.1 Let G be a super Lie group and $\mathcal{M}$ be a supermanifold. A super *principal G bundle* $\mathcal{P}(\mathcal{M}, G)$ is a triple $(\mathcal{P}, \pi, \mathcal{M})$, where $\mathcal{P}$ is a supermanifold and $\pi : E \to \mathcal{M}$ is a supersmooth map, which has the following properties:

(a) There is a free right action

$$\mathcal{P} \times G \to \mathcal{P}, \qquad (u, g) \to ug = R_g u \qquad (12.1)$$

of G on $\mathcal{P}$.

(b) The supermanifold $\mathcal{M}$ is the quotient of $\mathcal{P}$ under this group action, that is

$$\mathcal{M} = \mathcal{P}/G. \tag{12.2}$$

Also π is the canonical projection and is supersmooth.

(c) $\mathcal{P}$ is locally trivial, that is, there is an open cover $\{U_\alpha | \alpha \in \Lambda\}$ such that, for each $\alpha \in \Lambda$, there is a superdiffeomorphism $h_\alpha : \pi^{-1}(U_\alpha) \to U_\alpha \times G$ and a supersmooth map $\phi_\alpha : \pi^{-1}(U_\alpha) \to G$ with

$$h_\alpha(u) = (\pi(u), \phi_\alpha(u)) \quad \text{and} \quad \phi_\alpha(ug) = \phi_\alpha(u)g. \tag{12.3}$$

The space $\mathcal{M}$ is referred to as the base space, $\mathcal{P}$ as the total space and G as the structure group. Often the bundle will simply be referred to as $\mathcal{P}$. An immediate, if trivial, example is the cartesian product $\mathcal{M} \times G$ with G action $R_h(x,g) = (x, gh)$ and projection $\pi(x,g) = x$. It will be useful to define the transition functions of a super principal bundle $(\mathcal{P}, \pi, \mathcal{M})$, which are functions from the overlaps between trivialisation neighbourhoods into the structure group.

Definition 12.1.2   With the notation of Definition 12.1.1, suppose that α, β are elements of Λ such that $U_\alpha \cap U_\beta$ is non-empty. Then the *transition function* $g_{\alpha\beta}$ is defined by

$$g_{\alpha\beta} : U_\alpha \cap U_\beta \to G$$
$$u \mapsto \phi_\alpha(u) \, (\phi_\beta(u))^{-1}. \tag{12.4}$$

Given a super principal G bundle $\mathcal{P}$, corresponding to any left G space V there is an associated super bundle $\mathcal{P}(V)$ exactly as in the classical case; for completeness this will now be defined.

Definition 12.1.3   Let V be a left G space with G action $(g, v) \mapsto gv$ and $(\mathcal{P}, \pi, \mathcal{M})$ be a super principal G bundle. A right action of G on $\mathcal{P} \times V$ is defined by setting

$$(u, v)g = (ug, g^{-1}v). \tag{12.5}$$

The quotient of $\mathcal{P} \times V$ under this group action is denoted $\mathcal{P} \times_G V$; it consists of equivalence classes $[u, v]$ of points (u, v) in $\mathcal{P} \times V$ with $(u, v) \sim (u', v')$ if and only if there exists $g \in G$ such that $(u', v') = (ug, g^{-1}v)$. There is a well defined projection map $\pi_V : \mathcal{P} \times_G V \to \mathcal{M}$ defined by $\pi_V([u, v]) = \pi(u)$.

It may be seen that for each local trivialisation neighbourhood U_α of $\mathcal{M}$, the set $\pi_V^{-1}(U_\alpha)$ can be identified with the set $U_\alpha \times V$. This identification is used to put a supersmooth structure on $\mathcal{P} \times_G V$ in such a way that π_V is supersmooth. The triple $(\mathcal{P} \times_G \mathcal{M}, \pi_V, \mathcal{M})$ is the *associated super bundle* of $\mathcal{P}$ *via* V.

Both super principal super bundles and associated super bundles are examples of super fibre bundles, which are simply twisted products of a base space $\mathcal{M}$ and a fibre F, that is, triples $(E, \pi, \mathcal{M})$ with $\pi : E \to \mathcal{M}$ super-smooth and an open cover $\cup_\alpha U_\alpha = \mathcal{M}$ with the local trivialisation property

$$\pi^{-1}(U_\alpha) = U_\alpha \times F. \tag{12.6}$$

In the case of a super principal bundle the fibre is G while for an associated super bundle it is V. As in the classical case there is a notion of cross-section, associating with each point x in $\mathcal{M}$ exactly one point in the fibre $\pi^{-1}(x)$ over x.

Definition 12.1.4 A *cross section* of a super fibre bundle $(E, \pi, \mathcal{M})$ is a supersmooth mapping

$$s : \mathcal{M} \to E \tag{12.7}$$

such that $\pi \circ s$ is the identity on $\mathcal{M}$.

In most cases the fibre V of an associated super bundle will be a super vector space; if the dimension of V is (r, s) and the base space $\mathcal{M}$ is an (m, n)-dimensional supermanifold, then whenever U is an open subset of $\mathcal{M}$ which is both a coordinate neighbourhood of $\mathcal{M}$ and a local trivialisation neighbourhood for the bundle, any cross section of $\pi^{-1}(U)$ (the restriction of E to U) can be written in terms of local coordinates $(x; \xi)$ on $\mathcal{M}$ and a basis $(Y^1, \ldots, Y^{r+s})$ of V as

$$f(x; \xi) = \sum_{i=1}^{r+s} \sum_{\underline{\mu} \in \mathcal{M}_n} f_{\underline{\mu} i}(x) \xi^{\underline{\mu}} Y^i. \tag{12.8}$$

This local structure can be abstracted to characterise the sheaves used in the algebro-geometric formulation of super vector bundles as given for example in [100].

A connection can be defined on a super principal bundle in close analogue with the classical case.

Definition 12.1.5 Let $(\mathcal{P}, \pi, \mathcal{M})$ be a super principal G bundle. For each point u of $\mathcal{P}$ let $T_u\mathcal{P}$ denote the tangent space to $\mathcal{P}$ at u and let $V_u\mathcal{P}$ denote the kernel of $\pi_* : T_u\mathcal{P} \to T_{\pi(u)}\mathcal{M}$. Then a *connection* Γ on $(\mathcal{P}, \pi, \mathcal{M})$ is an assignment to each $u \in \mathcal{P}$ of a subspace $H_u\mathcal{P}$ of $T_u\mathcal{P}$ such that

(a) $T_u\mathcal{P} = V_u\mathcal{P} \oplus H_u\mathcal{P}$
(b) $H_{ug}\mathcal{P} = R_{g*}H_u\mathcal{P}$
(c) The dependence of $H_u\mathcal{P}$ on u is G^∞.

Each connection on $(\mathcal{P}, \pi, \mathcal{M})$ has a one-form ω associated with it taking values in the super Lie module of the structure group G. The definition uses the fundamental vector field $\underline{\xi}$ on $\mathcal{P}$ corresponding to $\xi \in \mathcal{L}(G)$, as in Definition 9.6.3.

Proposition 12.1.6 *Suppose that* $\omega : \mathcal{D}(\mathcal{P}) \to \mathcal{L}(G)$ *is defined by*

$$\omega(\underline{\xi}_u) = \xi \quad \text{for all} \quad \xi \in \mathcal{L}(G)$$
$$\omega(Y)(u) = 0 \quad \text{if} \quad Y_u \in H_u\mathcal{P}. \tag{12.9}$$

Then

(a) ω *is a one-form on* $\mathcal{P}$ *with values in* $\mathcal{L}(G)$,
(b) $R_g^*(\omega) = \mathrm{ad}(g^{-1})\omega$ *for all* g *in* G.

This one-form can be used to define covariant derivatives of cross sections of $\mathcal{P}$ and its associated bundles, and also a curvature two form, in the usual way.

12.2 The frame bundle and tensors

Associated with any (m, n)-dimensional supermanifold is its super principal $\mathrm{GL}(m, n; \mathbb{R}_S)$ bundle of linear frames, which will now be defined.

Definition 12.2.1 Let $\mathcal{M}$ be an (m, n)-dimensional supermanifold. A *linear frame* l at a point x in $\mathcal{M}$ is an ordered super basis $Y_1, \ldots, Y_{m+n}$ of the tangent space to $\mathcal{M}$ at x. Let $B(\mathcal{M})$ be the set of all linear frames at all points x of $\mathcal{M}$, and let $\pi : B(\mathcal{M}) \to \mathcal{M}$ be the mapping which takes a frame at x to the point x. Then $\mathrm{GL}(m, n; \mathbb{R}_S)$ acts on $B(\mathcal{M})$ on the right according to the rule

$$(Y_1, \ldots, Y_{r+s})g = (Y_1', \ldots, Y_{r+s}') \quad \text{with} \quad Y_i' = \sum_{k=1}^{m+n} Y_k g^k{}_i . \qquad (12.10)$$

$B(\mathcal{M})$ is given a $(m + m^2 + n^2, n + 2mn)$-dimensional differentiable struc-
ture by choosing an open cover $\{U_\alpha | \alpha \in \Lambda\}$ of $\mathcal{M}$ by coordinate neigh-
bourhoods. The collection $\{\pi^{-1}(U_\alpha) | \alpha \in \Lambda\}$ is then an open cover of
$B(\mathcal{M})$. If $(X_\alpha^1, \ldots, X^{m+n})$ are coordinates on U_α then the coordinates of
a frame $Y^1, \ldots, Y^{m+n}$ at a point x in U_α are defined to be $(X_\alpha^i, Y_{\alpha j}^i, i, j =
1, \ldots, r + s)$ where the matrix $(Y_{\alpha i}^j)$ is defined by

$$Y_i = \sum_{j=1}^{m+n} Y_{\alpha i}^j \partial_{X^j}^S . \qquad (12.11)$$

This also defines the local trivialisation of $B(\mathcal{M})$, and establishes that its
structure group is indeed $\mathrm{GL}(m, n; \mathbb{R}_S)$. The bundle $B(\mathcal{M})$ is called the
bundle of linear frames on $\mathcal{M}$.

Any representation of $\mathrm{GL}(m, n; \mathbb{R}_S)$ now defines an associated super
bundle of $B(\mathcal{M})$, leading to tensor bundles as in the classical case. A particular example, using the superdeterminant, leads to the notion of Berezin
density.

Definition 12.2.2 A *Berezinian density* on $\mathcal{M}$ is a cross section of the
bundle $\mathrm{Ber}(\mathcal{M})$ associated to the super frame bundle by the superdeterminant.

Some particular examples of tensors occur in the following sections. Given
a connection in the frame bundle, or any sub bundle, torsion can be defined
in the standard way.

12.3 Riemannian structures

It is possible to define an analogue of Riemann metric on a supermanifold.

Definition 12.3.1 A *Riemannian metric* on a supermanifold $\mathcal{M}$ is a
non-degenerate even graded symmetric 2-covariant tensor g on $\mathcal{M}$. (The
tensor is said to be non-degenerate if at each point p in $\mathcal{M}$ and for each
fixed Y_p in $T_p\mathcal{M}$

$$(Y_p, Z_p)g = 0 \, \forall Z_p \in T_p\mathcal{M} \quad \text{if and only if} \quad Y_p = 0 .) \qquad (12.12)$$

In a local coordinate system X^i a Riemann metric has components $g_{ij} = \left\langle \frac{\partial}{\partial X^i}, \frac{\partial}{\partial X^j} \middle| g \right\rangle$. If the dimension of $\mathcal{M}$ is (m, n) then the graded symmetry means that $g_{ij} = -g_{ji}$ if $i, j > m$ and $g_{ij} = g_{ji}$ otherwise. It is immediately obvious that a Riemannian metric can only exist on a supermanifold whose odd dimension is an even number.

A Riemannian metric on an $(m, 2l)$-dimensional supermanifold $\mathcal{M}$ is equivalent to a reduction of the frame bundle $B(\mathcal{M})$ to an $\mathrm{OSP}(m, l; \mathbb{R})$ bundle. This is seen by observing that the metric allows identification of preferred bases $\{E_A | A = 1, \ldots, m + 2l\}$ of the tangent space at each point which satisfy

$$\langle E_A, E_B | g \rangle = \Upsilon_{AB} \tag{12.13}$$

with Υ as in (9.13). In the superspace formulation of supergravity described in Section 13.2 the frame bundle is reduced to a smaller group. In either case the matrices $\left(E_M{}^A \right)$ corresponding to expansion of the basis $\left\{ E^A \right\}$ of one forms dual to the preferred basis $\{ E_A | A = 1, \ldots, m + 2l \}$ of the tangent space in terms of coordinate differentials $\mathrm{d}X^M$ as $E^A = E_M{}^A \mathrm{d}X^M$ provide local representatives $Z(X)$ of a Berezinian density Z which take the form

$$Z(X) = \mathrm{sdet} \left(E_A{}^M \right). \tag{12.14}$$

12.4   Even symplectic structures

On a supermanifold two kinds of symplectic structure have been considered, referred to as odd and even. Even symplectic structures, the subject of this section, make possible the extension to the super setting of many standard applications of symplectic structures. Both even and odd symplectic structures have proved useful in the quantization of theories with symmetry, as will be explained further in Chapter 16. In this section the definition of an even symplectic form is given, and some standard properties briefly described, including the Darboux theorem and, in the smooth case, a characterisation due to Rothstein[134].

Definition 12.4.1   An *even symplectic form* on a supermanifold $\mathcal{M}$ is an even 2-form ω which is closed and non-degenerate. An even symplectic supermanifold is a supermanifold $\mathcal{M}$ together with an even symplectic form ω on $\mathcal{M}$.

An even symplectic form is simply a graded-commutative analogue of a classical symplectic form, so that the Hamiltonian vector field associated to a smooth function can be defined in the following way:

Definition 12.4.2   Suppose that F is a G^∞ function on $\mathcal{M}$. Then a vector field Y_F of Grassmann parity $|F|$ is defined by

$$Y_F \omega^{-1}(\mathrm{d}F) \tag{12.15}$$

where ω here denotes the mapping from $T\mathcal{M}$ to $T^*\mathcal{M}$ defined by

$$\omega(Y)(Z) = \langle Y, Z | \omega \rangle . \tag{12.16}$$

(The vector field Y_F is referred to as the *Hamiltonian vector field* of F.)

The parity of the Hamiltonian vector field is the same as that of the corresponding function, and Poisson brackets (which are graded antisymmetric) may be defined as in the classical case.

A standard example of an even symplectic supermanifold (that is, a pair $(\mathcal{N}, \omega)$ with $\mathcal{N}$ a supermanifold and ω an even symplectic form on $\mathcal{N}$) is the cotangent space $T^*\mathcal{M}$ of an (m, n)-dimensional supermanifold $\mathcal{M}$, with

$$\omega = \sum_{i=1}^m \mathrm{d}p_i \wedge \mathrm{d}x^i + \sum_{j=1}^n \mathrm{d}\pi_j \wedge \mathrm{d}\xi^j \tag{12.17}$$

where the local coordinates of a one form α at a point q in $\mathcal{M}$ consist of local coordinates $(x; \xi)$ at q and the coefficients p_i, π_j in the expansion $\alpha = \sum_{i=1}^m p_i \mathrm{d}x^i + \sum_{j=1}^n \pi_j \mathrm{d}\xi^j$. There is also a Darboux theorem [95], which states that on a $(2l, n)$-dimensional symplectic supermanifold local coordinates $(x^i, p_i; \xi^j), i = 1, \dots l, j = 1, \dots, m$ may always be found in which the symplectic form ω takes the form

$$\omega = \sum_{i=1}^l \mathrm{d}p_i \wedge \mathrm{d}x^i + \sum_{j=1}^n \epsilon_j \, \mathrm{d}\xi^j \wedge \mathrm{d}\xi^j \tag{12.18}$$

with each ϵ_j equal to 1 or -1. An alternative standard form due to Rothstein will now be described.

In the supersmooth case Rothstein[134] has given a characterisation of symplectic structures on a supermanifold $\mathcal{M}$ in terms of classical data $(\mathcal{M}_{[\emptyset]}, \omega, E^*, g, \nabla)$ where ω is a symplectic structure on the body $\mathcal{M}_{[\emptyset]}$ of $\mathcal{M}$, E is the vector bundle corresponding to a choice of the splitting which must exist by Batchelor's theorem 8.2.1, g is a metric on $\mathcal{M}_{[\emptyset]}$ and ∇ is a compatible connection on E. To specify a symplectic form $\tilde{\omega}$ on $\mathcal{M}$ from this data, it is sufficient to specify $\tilde{\omega}(\widehat{X}, \widehat{Y})$, $\tilde{\omega}(\widehat{X}, \frac{\partial}{\partial \xi^j})$ and $\tilde{\omega}(\frac{\partial}{\partial \xi^i}, \frac{\partial}{\partial \xi^j})$ where X, Y are vector fields on $\mathcal{M}_{[\emptyset]}$ and $\xi^i, i = 1, \ldots, n$ are odd coordinates on $\mathcal{M}$ corresponding to the chosen splitting, and hence to a basis $\{e^i | i = 1, \ldots, n\}$ of E. The chosen symplectic structure is determined by setting

$$\tilde{\omega}(\widehat{X}, \widehat{Y}) = \omega(X, Y) + \tfrac{1}{2} R(X, Y)$$

$$\tilde{\omega}\left(\widehat{X}, \frac{\partial}{\partial \xi^j}\right) = 0$$

$$\text{and} \quad \tilde{\omega}\left(\frac{\partial}{\partial \xi^i}, \frac{\partial}{\partial \xi^j}\right) = g(e^i, e^j). \tag{12.19}$$

As well as showing that this does indeed define a symplectic form on $\mathcal{M}$, Rothstein also shows that any symplectic form on $\mathcal{M}$ can be expressed in this way[134].

12.5   Odd symplectic structures

Odd symplectic structures, which have no classical analogue, play an important role in the BV quantization scheme for theories with symmetry, as will be briefly explained in Chapter 16. The basic definition is very similar to that of an even symplectic structure.

Definition 12.5.1

(a) An *odd symplectic form* on a supermanifold $\mathcal{N}$ is an odd 2-form β which is closed and non-degenerate.
(b) An *odd symplectic supermanifold* is a pair $(\mathcal{N}, \beta)$ with $\mathcal{N}$ a supermanifold and β an odd symplectic form on $\mathcal{N}$.

An important theorem, due originally to Shander [140], gives the appropriate analogue of the Darboux theorem for odd symplectic supermanifolds:

Theorem 12.5.2   *Let $(\mathcal{N}, \beta)$ be an odd symplectic supermanifold. Then*

(a) the even and odd dimensions of $\mathcal{N}$ are equal, so that $\mathcal{N}$ has dimesnion (m, m) for some integer m;
(b) At every point of the (m, m)-dimensional supermanifold $\mathcal{N}$ there are local coordinates $(x^i; \xi^i), i = 1, \ldots, m$ such that

$$\beta = \sum_{i=1}^{m} \mathrm{d}\xi^i \wedge \mathrm{d}x^i. \tag{12.20}$$

A simple proof of this result was given by Khudaverdian [89]; just as the cotangent bundle provides an example of a classical symplectic manifold, the 'odd' cotangent bundle $S(M, T^*M)$ provides an example of an odd symplectic manifold, with

$$\beta = \sum_{i=1}^{m} \mathrm{d}\pi_i \wedge \mathrm{d}x^i \tag{12.21}$$

where x^i are local coordinates about a point q in M, and π_i are the odd coordinates corresponding to the basis $\mathrm{d}x^i$ of $T_q^* M$.

As in the even case, the non-degeneracy means that the symplectic form provides an invertible map from the tangent space to the cotangent space, and hence an analogue (with reverse parity) of the standard construction of the Hamiltonian vector field corresponding to a function on $\mathcal{N}$ can be defined. The corresponding analogue of the Poisson bracket plays an important role in BV quantization and is known as the antibracket. Further constructions involving odd symplectic structures may be found in the work of Khudaverdian [88, 89] and Khudaverdian and Voronov [90].

Chapter 13

Supermanifolds and supersymmetric theories

In quantum physics, a theory is said to possess supersymmetry if it possesses a symmetry which rotates fermionic degrees of freedom into bosonic and *vice versa*. Supersymmetric physics is a vast field, and this chapter is not intended to give a complete introduction to the subject, let alone a comprehensive list of references to the literature. What is covered in this chapter are some aspects of supersymmetry where mathematical aspects of supergeometry and superanalysis have been particularly significant, including superfield techniques, supergravity and super embeddings, while super Riemann surfaces (and hence string quantization) are discussed in Chapter 14 and path integrals on supermanifolds in Chapter 15. There is an extensive literature on supersymmetric physics, a useful account may be found in the book of West [160] and a clear introduction in the book of Freund [54]. Superspace methods in supersymmetric physics are described in considerable detail by Buchbinder and Kuzenko [28].

One of the first examples of a supersymmetric theory is the Wess-Zumino model [158], which was the first four-dimensional model with a linear realisation of supersymmetry. The fields of the theory are a scalar A, a pseudo scalar B, a Majorana spinor field χ and two auxiliary scalar fields F and G, all of which are functions of four-dimensional Minkowski space, and the action is

$$
\begin{aligned}
&S\left(A, B, \chi, F, G\right) \\
&= \int \mathrm{d}^4 x \left(-\tfrac{1}{2}\,\partial_m A \partial^m A - \tfrac{1}{2}\,\partial_m B \partial^m B - \tfrac{1}{2}\,\bar{\chi}\slashed{\partial}\chi + \tfrac{1}{2}\,F^2 + \tfrac{1}{2}\,G^2\right)
\end{aligned}
$$

$$(13.1)$$

for the massless free theory. (Mass and interaction terms are considered below.) Here the index and spinor conventions set down by West [160] have been used. Thus $m = 0, \ldots, 3$ is a four-dimensional Lorentz vector index and the conjugate spinor $\bar{\chi}$ of the Majorana spinor χ is defined by

$$\bar{\chi}^\alpha = \chi_\beta C^{\beta\alpha} \tag{13.2}$$

where C is the charge conjugation matrix and the Majorana spinor indices α, β take values from 1 to 4. The Dirac operator $\not{\partial}$ is defined by

$$\not{\partial} = \gamma_m \partial_m \tag{13.3}$$

where $\gamma_m, m = 0, \ldots, 3$ are Dirac matrices which satisfy

$$\gamma_m \gamma_n + \gamma_n \gamma_m = 2\,\eta_{mn} \tag{13.4}$$

with η_{mn} the Minkowski metric. The action is a Lorentz scalar. The fields F and G, which do not propagate, are known as *auxiliary fields* and are necessary for the multiplet of fields to carry a representation of the super-symmetry algebra; if they are excluded the algebra of the supersymmetry transformations on the remaining fields only closes when the equations of motion are satisfied, a situation referred to as *on-shell* supersymmetry.

In the original description of this model, the symmetries were expressed in terms of infinitesimal generators $P_m, m = 0, \ldots, 3$ and $Q_\alpha, \alpha = 1, \ldots, 4$ obeying a super Lie algebra

$$[P_m, P_n] = 0$$
$$[P_m, Q_\alpha] = 0$$
$$[Q_\alpha, Q_\beta] = 2\,(\gamma_m C)_{\alpha\beta}\, P^m \tag{13.5}$$

with each P_m even and each Q_α odd. The infinitesimal parameter a^m for the each P_m transformation was considered to be a commuting number while the parameter ϵ^α of each Q_α transformation was considered to be part of an anticommuting set so that

$$\epsilon^\alpha \epsilon^\beta = -\epsilon^\beta \epsilon^\alpha$$
$$\text{while}\quad \epsilon^\alpha a^m = a^m \epsilon^\alpha$$
$$\text{and}\quad a^m a^n = a^n a^m\,. \tag{13.6}$$

In other words, the parameters were taken to be even and odd elements of a super commutative algebra. The components of the spinor χ were also taken to be anticommuting. Combining parameters and generators to give an infinitesimal transformation of the form $a^m P_m + \epsilon^\alpha Q_\alpha$ gave elements of a Lie algebra, and so demonstrated that the infinitesimal transformations could be exponentiated to form a group acting on the fields; thus, albeit in a somewhat heuristic way, the concept of a super Lie group emerged.

The P_m are defined to be generators of translations, while the action of the infinitesimal supersymmetry transformation $\epsilon^\alpha Q_\alpha$ on the fields of the Wess-Zumino model is given by

$$
\begin{aligned}
\delta_\epsilon A &= \bar{\epsilon}\chi \\
\delta_\epsilon B &= i\bar{\epsilon}\gamma_5\chi \\
\delta_\epsilon \chi &= F + i\gamma_5 G + \partial\!\!\!/\,(A + i\gamma_5 B)\,\epsilon \\
\delta_\epsilon F &= \bar{\epsilon}\partial\!\!\!/\,\chi \\
\delta_\epsilon G &= i\bar{\epsilon}\gamma_5\partial\!\!\!/\,\chi\,.
\end{aligned}
\tag{13.7}
$$

It can be shown by explicit calculation that the action (13.1) is invariant under these transformation; also the action is manifestly invariant under translations and Lorentz rotations. (The invariance of the action can also be demonstrated by superspace techniques, using the form of the action (13.25).) Equipped with concepts from supermanifold theory, it can be seen that that the fields can be regarded as maps from $\mathbb{R}^4 \to \mathbb{R}_S^{4,4}$ with

$$
x \mapsto (A(x), B(x), F(x), G(x); \chi_1, \chi_2, \chi_3, \chi_4)
\tag{13.8}
$$

and the supersymmetry to be a symmetry under the action of the super Poincaré group described in Section 9.2.

Many other supersymmetric models have been constructed and analysed, both in flat Minkowski spacetime as described here and in curved spacetime as described below, where theories of supergravity are obtained. Theories such as that described above are said to have $N = 1$ supersymmetry, the number N specifying the number of spinor generators Q. In four dimensions N may be as high as 8. It is also possible to vary the dimension of the theory. Supersymmetric quantum mechanics, a theory in one spacetime dimension, has important geometric applications, and will appear in Chapters 15, 16 and 17. A more general book on this subject is [33]. Supergravity in two dimensions is important in string theory, and leads to the interesting concept of super Riemann surface considered in Chapter 14. A more recent variant is the introduction of a second kind of anticommuting Majorana spinor by Kleppe and Wainwright [91] which allows a new model based on the super sphere.

In fact in supersymmetric physics applications are found of maps between supermanifolds of a wide variety of dimension; for instance in string theory for the standard covariant Green Schwarz formalism of the superstring the target is a supermanifold but the worldsheet is purely bosonic, while in the formulation of the Neveu Schwarz Ramond spinning string of Friedan, Martinec and Shenker [57] the world sheet is a super Riemann surface. The two approaches were first combined by Berkovits [21], leading to many new sigma models on coset supermanifolds, for instance as in [22]. These in turn involve quite subtle features of representations of super Lie groups, as may be seen for instance in [63].

The superfield approach is described in Section 13.1, and extended to supergravity in Section 13.2. In Section 13.3 the superembedding technique for anaysing extended supersymmetric objects is discussed.

13.1 Superfields and the superspace formalism

The idea of superspace, which first appeared in papers of Salam and Strathdee [135] and Volkov and Akulov [152], comes from the observation that the 4-dimensional super Poincaré group can be realised as a group of transformations of a space with 4 commuting variables and 4 anticommuting variables. Recalling from Section 9.2 that the super Poincaré group is the semi direct product of the supertranslation group and the spin double cover $SL(2, \mathbb{C})$ of the Lorentz group $SO(1,3)$, if $(x; \theta)$ is an element of $\mathbb{R}_S^{4,4}$ with x^i even, $i = 0, \ldots, 3$ and θ^α odd, $\alpha = 1, \ldots, 4$, x is defined to transform as an $SO(1,3)$ vector and θ as a Majorana spinor. The action of the super translation part of the group is defined, for $(y, \epsilon) \in T^{4,4}$, by

$$(y, \epsilon) : \mathbb{R}_S^{4,4} \to \mathbb{R}_S^{4,4}$$
$$(x^m; \theta) \mapsto (x^m + y^m - \bar{\epsilon}^\alpha \gamma^m{}_{\alpha\beta} \theta^\beta, \theta + \epsilon), \qquad (13.9)$$

which is simply the action of the supertranslation group $T^{4,4}$ on itself if the standard identification of $T^{4,4}$ with $\mathbb{R}_S^{4,4}$ is made. This inspired the idea that one should consider *superfields*, that is, fields which are functions of superspace rather than simply of space.

A superfield is then a function f on $\mathbb{R}_S^{4,4}$ and thus has a Taylor expansion in the θ^α with

$$f(x; \theta) = f_{[\emptyset]}(x) + f_\alpha \theta^\alpha + \ldots . \qquad (13.10)$$

The components $f_{[\emptyset]}, f_\alpha$ and so on contain the physical fields, whose supersymmetry transformations are immediately determined by the action of the supersymmetry group on superspace, by the simple rule that

$$g(f)(x;\theta) = f(g(x;\theta)) \qquad (13.11)$$

where g denotes an element of the supersymmetry group and $g(x;\theta)$ is the image of the point $(x;\theta)$ under the action of this group. Of course there is no guarantee that the representation of the supersymmetry group corresponding to a superfield is irreducible, and indeed a simple superfield of the kind just described is in fact highly reducible. Because the odd coordinates θ^α are in a representation of the spin group $\mathrm{SL}(2,\mathbb{C})$, decompositions of $\mathrm{SL}(2,\mathbb{C})$ representations allow one to identify the spin content of the component fields. In general the superfield itself can be a multi-component object, transforming under some representation of the Lorentz group.

Two important features of the superspace formalism are that it provides a mechanism for constructing representations of the supersymmetry group, and hence of the super Poincaré group, and that it leads to supersymmetry invariants. In principle this allows a quantization procedure where by working in superspace supersymmetry is manifestly maintained, and the benefits of supersymmetry readily harvested. In practice the situation tends to be more complicated.

To illustrate these ideas, the Wess Zumino model will be constructed in superspace. The starting point is a superfield of the form (13.10). Simple counting immediately shows that the component fields are too many in number and with the wrong spin content to correspond exactly with the fields of the theory. In order to obtain the correct representation of the supersymmetry group it is necessary to restrict the superfield.

At this stage it is useful to introduce two-component spinors, with $\chi^\alpha = (\chi^A, \zeta_{\dot{A}})$ where

$$\chi^A = \tfrac{1}{2}(1 + \gamma_5)\chi \,, \qquad \zeta_{\dot{A}} = \tfrac{1}{2}(1 - \gamma_5)\chi \,, \qquad (13.12)$$

with, again using the conventions of West [160],

$$\gamma^m = \begin{pmatrix} 0 & \sigma^m \\ i\bar{\sigma}^m & 0 \end{pmatrix} \qquad (13.13)$$

where $\sigma^i, i = 1, 2, 3$ are Pauli matrices and σ^0 is the identity.

In the two-component spinor notation the odd coordinates on superspace are $(\theta^A, \theta_{\dot{A}})$ with

$$\theta^A = \tfrac{1}{2}(1 + \gamma_5)\theta \,, \qquad \theta_{\dot{A}} = \tfrac{1}{2}(1 - \gamma_5)\theta \,, \qquad (13.14)$$

where the same letter can be used for the undotted and dotted parts in view of the Majorana condition. There are thus two complex rather than four real odd coordinates. (This is similar to using coordinates $z, \bar{z}$ on $\mathbb{R}^2$.) The supersymmetry algebra now takes the form

$$[P_m, P_n] = 0$$
$$\left[P_m, Q^A\right] = [P_m, Q_{\dot{A}}] = 0$$
$$\left[Q^A, Q_{\dot{B}}\right] = 2(\sigma_m)^A{}_{\dot{B}} P^m$$
$$\left[Q^A, Q^B\right] = [Q_{\dot{A}}, Q_{\dot{B}}] = 0 \,. \tag{13.15}$$

The action of these generators on a superfield is given by

$$P_m = \frac{\partial}{\partial x_m}$$

$$Q_A = \frac{\partial}{\partial \theta^A} + i\theta^{\dot{A}}(\sigma^m)_{A\dot{A}} P_m$$

$$Q_{\dot{A}} = \frac{\partial}{\partial \theta^{\dot{A}}} + i\theta^A(\sigma^m)_{A\dot{A}} P_m \,. \tag{13.16}$$

This representation of these operators allows one to see that there are odd first-order differential operators D_A and $D_{\dot{A}}$ which commute with the supersymmetry group. These operators are

$$D_A = \frac{\partial}{\partial \theta^A} - i\theta^{\dot{A}}(\sigma^m)_{A\dot{A}} P_m \,,$$

$$D_{\dot{A}} = \frac{\partial}{\partial \theta^{\dot{A}}} - i\theta^A(\sigma^m)_{A\dot{A}} P_m \,. \tag{13.17}$$

Direct calculation shows that both D_A and $D_{\dot{A}}$ commute with P_m, Q_A and $Q_{\dot{A}}$. It will be useful to note that

$$\left[D^A, D_{\dot{B}}\right] = 2(\sigma_m)^A{}_{\dot{B}} P^m \,. \tag{13.18}$$

Because $D_{\dot{A}}$ commutes with the supersymmetry generators, a superfield f which satisfies the constraint

$$D_{\dot{A}} f = 0 \tag{13.19}$$

will provide a representation of the supersymmetry group. (Such superfields are referred to as *chiral superfields*.) This representation can be analysed by setting

$$a = f(x,\theta)|_{\theta=0}, \quad \rho_A = D_A f(x,\theta)|_{\theta=0} \quad \text{and} \quad y = -\frac{1}{2} D^A D_A f(x,\theta)|_{\theta=0}.$$
(13.20)

No further independent component field can exist because of (13.18) and (13.19). The transformations of these fields can be evaluated using the explicit form (13.16) of the supersymmetry generators. This shows that the fields of the Wess-Zumino model can be identified with those of the constrained superfield f by taking

$$a = -\frac{1}{2}(A + iB), \quad \chi^A = \rho^A \quad \text{and} \quad y = F + iG.$$
(13.21)

The integration theory of Chapter 11 provides a method for constructing supersymmetry invariants from superfields. The simplest way to do this is to observe that, since supersymmetry transformations are implemented on superfields by super translations in superspace, the integral combinations of superfields is invariant under a supersymmetry transformation provided that all superfields vanish on the boundary of the region of integration, or fall away sufficiently fast at infinity. There are further methods, corresponding to integrating over subspaces or quotient spaces of superspace, which lead to invariants constructed from constrained superfields.

A useful observation is to note that, provided the integrand obeys suitable boundary conditions, an integral over superspace can be represented in terms of an ordinary integral over spacetime combined with super derivatives $D_A, D_{\dot{A}}$ in a number of ways. This is because the Berezin integral is equivalent to a series of odd derivatives, so that

$$\int \mathrm{d}^4 x \mathrm{d}^4 \theta f(x,\theta) = \int \mathrm{d}^4 x \frac{\partial}{\partial \theta^{\dot{2}}} \frac{\partial}{\partial \theta^{\dot{2}}} \frac{\partial}{\partial \theta^2} \frac{\partial}{\partial \theta^1} f(x,\theta)$$
(13.22)

and the odd derivatives $\frac{\partial}{\partial \theta^A}, \frac{\partial}{\partial \theta^{\dot{A}}}$ can be replaced by the corresponding super derivatives $D_A, D_{\dot{A}}$ in cases where the difference is a boundary term which is assumed to vanish. In particular

$$\int \mathrm{d}^4 x \mathrm{d}^4 \theta f(x,\theta) = \int \mathrm{d}^4 x D^A D_A D^{\dot{A}} D_{\dot{A}} f(x,\theta).$$
(13.23)

This of course shows that the integral over the full $(4,4)$-dimensional superspace of any combination of chiral superfields (which satisfy the constraint (13.19)) is zero. In order to obtain invariant combinations of chiral superfields a $(4,2)$-dimensional superspace $\mathcal{S}$, the quotient of $\mathbb{R}^{4,4}_S$ by the

$(0,2)$-dimensional super Lie group generated by $\left\{ D_{\dot{A}} | \dot{A} = 1, 2 \right\}$, must be used. As a supermanifold this is superdiffeomorphic to $\mathbb{R}_S^{4,2}$, so that integration on this space is well-defined. Also any chiral superfield on the full superspace $\mathbb{R}_S^{4,4}$ induces a well defined function on $\mathcal{S}$, so that an integral

$$\int_{\mathcal{S}} \mathrm{d}^4 x \mathrm{d}^2 \theta f(x; \theta)$$

will be a supersymmetry invariant. This integral satisfies

$$\int_{\mathcal{S}} \mathrm{d}^4 x \mathrm{d}^2 \theta f(x; \theta) = \int \mathrm{d}^4 x D^A D_A f(x, \theta) \,. \tag{13.24}$$

Of course as well as the chiral constraint (13.19), there is the complex conjugate constraint $D_A = 0$, with an analogous procedure for constructing invariants. An example of an invariant constructed by these methods is the kinetic term in the action for the Wess Zumino model. In terms of component fields this has the form (13.1); this is equal to the superspace integral

$$\int \mathrm{d}^4 x \mathrm{d}^4 \theta f(x, \theta) \bar{f}(x; \theta) \tag{13.25}$$

with component fields identified as in (13.20). The mass term takes the form

$$\int_{\mathcal{S}} \mathrm{d}^4 x \mathrm{d}^2 \theta \; m(f(x; \theta))^2 + \text{complex conjugate} \,, \tag{13.26}$$

and interaction terms can be built in a similar manner, for instance

$$\int_{\mathcal{S}} \mathrm{d}^4 x \mathrm{d}^2 \theta \; \lambda(f(x; \theta))^3 + \text{complex conjugate} \,. \tag{13.27}$$

In this section a simple example has been described to illustrate the superfield formalism for a supersymmetric theory. There has been no attempt at a comprehensive treatment of this vast subject; a much more complete account can be found in the book of Buchbinder and Kuzenko [28], while many standard texts on supersymmetry, such as West [160], include an account of superspace methods. In the next section these superspace constructions will be extended to the geometric setting of general relativity, so that theories of supergravity can be included. One of the principal reasons for introducing superfield techniques is that they allow quantization methods to be cast in a manifestly supersymmetric form.

A key motivations for studying supersymmetric theories is that they are expected to have much better quantum properties, with cancellations of the infinities which plague many quantum field theories; superfield techniques such as super functional integration and super Feynman diagrams have allowed quantum calculations which show these hopes to be realised in some cases. A pioneering example was the work of Salam and Strathdee [135] in the case of rigid $N = 1$ supersymmetry. These ideas provide a good illustration of the power of super mathematics.

13.2 Supergravity

Supergravity theories have local supersymmetry, as opposed to the global or rigid supersymmetry of the Wess Zumino described in the preceding section. The first examples of such theories were written down by Ferrara, Freedman and van Nieuwenhuizen [50] and by Deser and Zumino [40]; the spin 2 gravity field e^a_m appears in a super multiplet which also contains a spin $\frac{3}{2}$ Rarita-Schwinger field ψ. These models have on-shell supersymmetry, the model containing the minimal auxiliary fields for $N = 1$ supergravity was constructed by Stelle and West [146] and by Ferrara and van Nieuwenhuizen [51]. In this section the construction of this model in superspace will be described as an illustration of the supermanifold techniques used in uniting supersymmetry with gravity.

The first attempt at a superspace version of supergravity was understandably based on the idea that the appropriate geometry would be Riemannian geometry on a supermanifold [2]. This corresponds to a reduction of the frame bundle of the (m, n) supermanifold concerned to a $\mathrm{OSP}(m, n)$ bundle, which is a larger group than the rotation part of the corresponding rigid superspace theory. Just as gravity uses a geometrical structure which allows local Lorentz frames which carry an affine representation of the Poincaré group, the appropriate supergeometry, as first realised by Wess and Zumino [159], must allow local frames carrying a representation of the super Poincaré group. Thus the starting point in constructing supergravity in superspace is to consider the spin bundle of the spacetime manifold concerned. The model considered by Wess and Zumino was based on four spacetime dimensions; the 4-dimensional spacetime of the theory will be denoted P, the metric g and the spin bundle spinP, while the 4-dimensional vector bundle over P associated to this spin bundle via the Majorana representation of the spin group $\mathrm{SL}(2, \mathbb{C})$ will be denoted S. The required

$(4, 4)$-dimensional supermanifold is then $\tilde{P} = S(P, S)$. As coordinate neighbourhoods for this supermanifold contractible coordinate neighbourhoods of P may be taken, which will also be trivialisation neighbourhoods for the spin bundle, obtaining local coordinates $(x^m; \theta^\alpha)$ with transition functions taking the standard form (8.5) so that the even coordinates x^m transform as on P while the odd coordinates θ^α rotate according to the Majorana representation of the transition functions of the spin bundle. This will provide us with an atlas of charts on $\tilde{P}$, but more general coordinate systems $(x^m; \theta^\mu)$ will also be needed. The Riemannian metric on P, together with the particular construction of the supermanifold $\check{M}$, means that the frame bundle of $\tilde{P}$ admits a reduction to a $SL(2, \mathbb{C})$ bundle; elements of this bundle are known as *vielbein* and denoted $(E_\mathcal{A}) = (E_a, E_\alpha) = (E_a, E_A, E_{\dot{A}})$ where a is an $SO(3, 1)$ vector index and α and $A, \dot{A}$ are 4 and 2 component spinor indices as before, while $\mathcal{A}$ includes both vector and spinor indices. (Here the convention used is that letters near the beginning of the alphabet are used for indices in the vielbein basis, while letters in the middle of the alphabet are used for coordinate indices. Caligraphic capitals are used for general indices, lower case Latin indices for vector indices, lower case Greek for 4 component spinors and plain capitals for 2 component spinors.)

Elements of the vielbein bundle are ordered bases of the tangent space at points of $\check{M}$, and so can be expanded in a general coordinate basis $\frac{\partial}{\partial X^\mathcal{M}}$ giving a vielbein matrix

$$\left(E_\mathcal{A}^{\ \mathcal{M}}\right) = \begin{pmatrix} E_a^{\ m} & E_a^{\ \mu} \\ E_\alpha^{\ m} & E_\alpha^{\ \mu} \end{pmatrix}. \tag{13.28}$$

It is also useful to consider the dual basis $E^\mathcal{A}$ of the cotangent space and the corresponding inverse vielbein matrices

$$\left(E_\mathcal{M}^{\ \mathcal{A}}\right) = \begin{pmatrix} E_m^{\ a} & E_m^{\ \alpha} \\ E_\mu^{\ a} & E_\mu^{\ \alpha} \end{pmatrix} \tag{13.29}$$

corresponding to expansion in the basis $dX^\mathcal{M}$ of coordinate differentials. The inverse vielbein matrix elements contain the fields of the system. In order to obtain the correct field content, implement the supersymmetry transformations and construct the action of the theory further geometry is required. A connection ϕ in the reduced frame bundle is required whose torsion, while not required to be zero as in conventional Riemannian geometry, is still subject to some constraints. Working in a vielbein basis the connection has components $\phi_{\mathcal{A}a}^{\ \ b}$ and $\phi_{\mathcal{A}\alpha}^{\ \ \beta}$. Wess and Zumino specified that certain torsion components should satisfy constraints,

$$T_{\alpha\beta}{}^{c} = \gamma_{\alpha\beta}^{c}, \quad T_{\alpha\beta}{}^{\gamma} = 0, \quad T_{ab}{}^{c} = T_{a\beta}{}^{c} = 0 \quad \text{and} \quad T_{ab}{}^{c} = 0, \quad (13.30)$$

while the other components are unconstrained. Some aspects of these constraints follow from geometric consistency and the requirement that rigid supersymmetry should emerge in the limit where the vielbein matrix is the identity matrix, others are justified by the physical content of the resulting theory. Identifying the physical fields and calculating the supersymmetry transformations is a technical procedure using the Bianchi identities, whose properties were analysed by Dragon [46]. Details of this process are beyond the scope of this book, and may be found in a number of texts including [28, 157, 160].

Because the superspace formulation of supergravity includes vielbein, it is again possible to use integration to construct supersymmetry invariants. This is because the vielbein provide a Berezin density (as in Section 12.3) which takes the form

$$E = \text{sdet}\left(E_{\mathcal{M}}{}^{\mathcal{A}}\right) \, \mathrm{d}^{4}x \, \mathrm{d}^{4}\theta. \quad (13.31)$$

A simple example of such an invariant is

$$\int E = \text{sdet}\left(E_{\mathcal{M}}{}^{\mathcal{A}}\right) \, \mathrm{d}^{4}x \, \mathrm{d}^{4}\theta. \quad (13.32)$$

which, if multiplied by the correct factor, is in fact the action for the $N = 1$ supergravity theory with minimal auxiliary fields.

There are many other possible supergravity theories beyond the four-dimensional model discussed here; both the dimension and the number of supersymmetries may be increased. For theories in d spacetime dimensions with N supersymmetries the supermanifold used usually has odd dimension $2^{[d]/2}N$, since $2^{[d]/2}$ is the dimension of the spinor representation in d dimensions. (Exceptions occur for instance when Majorana Weyl spinors exist.) The supermanifold is constructed in the standard way from the Nth power of the spinor bundle, and carries vielbein $E_{\mathcal{M}}{}^{\mathcal{A}}$ corresponding when $N = 1$ to a reduction of the bundle of super frames to the spin group as above. When $N > 1$ the reduction is more complicated. Super manifolds with these structures will be said to carry *supergravity geometry*. Each supergravity model requires its own set of torsion constraints; models exist whose superspace formulation is incomplete,e there is no universal prescription for constructing the supergeometry. Various approaches have been adopted to overcome these difficulties, particularly interesting from the point of view of supermanifold geometry being the harmonic superspace of Galperin, Ivanov, Ogievetsky and Sokatchev[59]. There are also problems with constructing actions and other necessary invariants as superspace integrals, because of the physical dimensions of such objects. However where the superspace formalism is complete it provides a powerful tool for maintaining supersymmetry during quantization and analyzing possible counterterms and hence renormalisability.

In two dimensions the Hilbert action for gravity is a topological invariant; while such models do not lead to propagating fields, they have proved to be of importance both in topological quantum field theory and in string theory. Two-dimensional supergravity is the basis of the Brink, Vecchia and Howe action for the spinning string [27], which is the supersymmetric version of the Polyakov action [111] for the bosonic string. The geometry of the superspace formulation of two dimensional supergravity is considered in Chapter 14 where, combined with superconformal transformations, it leads to the notion of super Riemann surface.

13.3   Super embeddings

In this section a super geometry technique, referred to as *superembedding*, for the study of supersymmetric extended objects know as branes is described. This approach was first developed by Sorokin, Tkach, Volkov and Zheltukhin [142, 143] for super particles, and then developed by a number of authors for extended objects. A unifying picture, using the de Rham cohomology of supermanifolds, was presented by Howe, Raetzel and Sezgin in [77], and it is the generic, supergeometric features of their approach which is described here.

In general a p-brane is an p-dimensional object moving in a D dimensional spacetime, sweeping out a $(p+1)$-dimensional subspace known as the world surface. A 0-brane is a particle, a 2-brane a string and so on. The first steps towards such theories of embedded objects came with string theory; a bosonic string sweeps out a two dimensional surface in spacetime. There are two ways in which string theory can acquire supersymmetry. The first is the so called *spinning string* which is considered further in Chapter 14. This can be formulated by making the world surface supersymmetric, leading to a super world surface which is a $(2,2)$-dimensional supermanifold. The second way in which a string can acquire supersymmetry is when the spacetime in which the string moves becomes a superspace, which leads to Green-Schwarz superstrings [64]. The work of Sorokin, Tkach, Volkov and Zheltukhin [142, 143] began a unification of these approaches, providing a natural geometric explanation of the κ or Siegel symmetry of the Green-Schwarz super string (and of the super particle). At the same time theories of p-branes with higher values of p were being developed, such as the Dirichlet or D-branes on which open strings have their ends. The superembedding formalism of [77] handles supersymmetric p-branes for all known values of p in a formalism where both the world surface and the space time are supermanifolds.

A *super embedding* is an embedding $f : \mathcal{M} \to \underline{\mathcal{M}}$ where both $\mathcal{M}$ and $\underline{\mathcal{M}}$ carry supergravity geometry with vielbein $E^{\mathcal{A}}$ and $E^{\underline{\mathcal{A}}}$ respectively; the domain $\mathcal{M}$ is known as the world surface, and has dimension $(d, \frac{1}{2}D')$ while the target space $\underline{\mathcal{M}}$ has dimension (D, D'). Local coordinates on $\mathcal{M}$ are denoted by $X = (x; \xi)$ and on $\underline{\mathcal{M}}$ by $\underline{X} = (\underline{x}, \underline{\xi})$ so that in local coordinates the super embedding is expressed as $\underline{Z}(Z)$. It is useful to define the embedding matrix

$$\mathcal{E}_{\mathcal{A}}{}^{\underline{\mathcal{A}}} = E_{\mathcal{A}}{}^{M} \frac{\partial X^{\underline{M}}}{\partial X^{M}} E_{\underline{M}}{}^{\underline{\mathcal{A}}}. \tag{13.33}$$

The embedding matrix is thus the differential of the embedding map referred to the preferred bases $E_{\mathcal{A}}$ and $E^{\underline{\mathcal{A}}}$ of the tangent space of $\mathcal{M}$ and the cotangent space of $\underline{\mathcal{M}}$ respectively.

A key geometric features of the superembedding method is the *embedding condition* [78] which requires that the odd tangent bundle of the world surface should be a sub-bundle of the odd tangent bundle of the target super spacetime. This embedding condition can be expressed in terms of the embedding matrix as

$$\mathcal{E}_{\alpha}{}^{\underline{a}} = 0 \tag{13.34}$$

where as before α is a spinor index and a is a Lorentz vector index. This condition allows the identification of supersymmetry multiplets for the theory concerned, along with their transformation properties. Details of this procedure, which are beyond the scope of this book, may be found in [77]. As with the torsion constraints in the superspace formulation of supergravity, the justification for this embedding condition is partly from the physics of the theory concerned.

It will now be shown that the de Rham theorem for smooth supermanifolds allows actions for p-branes to be constructed. This method, which was developed by Howe, Raetzel and Sezgin [77], uses another key feature

of the superembedding formalism for a brane, its Wess-Zumino form. For
a p-brane this is a closed $p + 2$ exterior form W on the super world surface
$\mathcal{M}$. Its particular form varies from theory to theory. Since W is closed it
is a representative of a $p + 2$ de Rham cohomology class on the G^∞ supermanifold $\mathcal{M}$. However by Theorem 10.3.5 the de Rham cohomology of $\mathcal{M}$
is isomorphic to that of its body $\mathcal{M}_{[\emptyset]}$ which is a $(p + 1)$-dimensional C^∞
manifold and thus has no nontrivial de Rham cohomology at degree higher
than $p + 1$, and so the Wess-Zumino form W must be exact. As a result
there will exist a $p + 1$ exterior L form on $\mathcal{M}$ which satisfies

$$\mathrm{d}L = W . \tag{13.35}$$

The $p+1$ exterior form L is unique up to the exterior derivative of a p-form,
and so the integral

$$S = \int_{\mathcal{M}_{[\emptyset]}} L \tag{13.36}$$

(constructed as in Section 11.7) provides a function of the fields of the
theory which is invariant under supersymmetry transformations. In many
cases of interest this integral is the action which determines the dynamics.

Chapter 14

Super Riemann surfaces

A super Riemann surface is a $(1,1)$-dimensional complex supermanifold which obeys an extra superconformal condition which is described below. From the point of view of supermanifold theory super Riemann surfaces are interesting because they provide the simplest examples of non-split supermanifolds. However the main motivation for the study of super Riemann surfaces has come from string theory, since they arise naturally in the extension of the Polyakov quantization method to the spinning string using the approach of Brink, P. Di Vecchia and Howe [27].

The notion of super Riemann surface follows from Howe's analysis [76] of the superspace geometry of superconformal transformations in two dimensions, and was first defined by Baranov and Schwarz [6, 7] and by Friedan [56] in their work on string theory. Much of the basic theory of super Riemann surfaces may be found in the paper of Crane and Rabin [37], while many other developments of the theory of super Riemann surfaces and its application to the quantization of the spinning string were given by Rosly and Schwarz and Voronov [130, 131], Baranov, Manin, Frolov and Schwarz [5] and Belavin and Knizhnik [15]. While it is possible to describe super Riemann surfaces in the algebro-geometric language, the data of an individual super Riemann surface includes odd moduli, so that either some auxiliary algebra is required or families of super Riemann surfaces must be studied as in the work of LeBrun and Rothstein [98]. Further developments of the theory of super Riemann surfaces have been made by Freund and Rabin[55], who proved that all super tori are algebraic, and by Rabin and Topiwala[115] who extended this result to all super Riemann surfaces. Rabin has also studied super elliptic curves[113], while Myung has introduced theta functions on such super Riemann surfaces[106]. In this book we confine attention to $N = 1$ super Riemann surfaces. Extended super Riemann surfaces have been studied by Cohn[31].

Section 14.1 describes Howe's formulation of the geometry of the spinning string in superspace, which is the basis of the definition of super Riemann surface given in Section 14.2. Section 14.3 considers the supermoduli space of super Riemann surfaces, and in Section 14.4 a theory of super contour integration is developed. Finally in Section 14.5 the algebraic structure of various spaces of functions on super Riemann surfaces is discussed.

14.1   The superspace geometry of the spinning string

The starting point of the superspace quantization of the spinning string is a Euclidean version of Howe's formulation of the classical dynamics in $(2,2)$-dimensional superspace [76]. This theory is constructed on the standard $(2,2)$-dimensional supermanifold for $N = 1$ supersymmetry constructed over a 2-dimensional Riemannian manifold with spin structure described in Section 13.2. Of course this supermanifold, which is a real, G^∞ supermanifold, is a split supermanifold, it is the superconformal features introduced below which lead in general to a non-split complex supermanifold structure. The classical action for this theory is

$$S = \tfrac{1}{4} \int \mathrm{d}^2 x\, \mathrm{d}^2\theta\; E\, \mathcal{D}_\alpha V\, \mathcal{D}^\alpha V \qquad (14.1)$$

where V is a supersmooth function on $\mathcal{M}$.

The construction of the $(2,2)$-dimensional supermanifold $\mathcal{M}$ makes it clear that there is a reduction of the super frame bundle of $\mathcal{M}$ from the super Lie group $\mathrm{GL}(2,2;\mathbb{R}_S)$ to the group $\mathrm{U}(1)$. This reduced bundle is central to Howe's construction, and it is useful to use vielbein (c.f. Section 13.2) which encode this explicitly, in other words, to select a super basis $\{E_a, E_\alpha | a = 1, 2, \alpha = 1, 2\}$ of the tangent space to $\mathcal{M}$ which transform under $e^{it} \in \mathrm{U}(1)$ according to the rule

$$\begin{pmatrix} E_a \\ E_\alpha \end{pmatrix} \mapsto \begin{pmatrix} V(e^{it}) & 0 \\ 0 & S(e^{it}) \end{pmatrix} \begin{pmatrix} E_a \\ E_\alpha \end{pmatrix} \qquad (14.2)$$

where V and S are respectively vector and spinor representations of $\mathrm{U}(1)$. Already this tight structure makes it clear that one is not simply considering Riemannian geometry on a supermanifold. As in the case of higher dimensional supergravity theories, further structure is required to generate the correct physical theory. The next step is to choose a connection in the vielbein bundle such that the torsion $T_{AB}{}^C$ satisfies the constraints

$$T_{\alpha\beta}{}^c = \gamma_{\alpha\beta}^c, \qquad T_{\alpha\beta}{}^\gamma = 0 \quad \text{and} \quad T_{ab}{}^c = 0. \tag{14.3}$$

These constraints mean that the various components $E_A{}^{\mathcal{M}}$ of the vielbein cannot be freely specified; it is shown by Howe [76] that these constraints do lead to the desired physical theory of the spinning string. Although the $(2,2)$-dimensional geometry may seem a little messy it is essential as a bridge between the 'real' world of physics and the mathematically elegant world of super Riemann surfaces. Polyakov-Brink-Di Vecchia-Howe quantization involves integration over the space of all $E_M{}^A$ which satisfy the constraints and all fields V, modulo the symmetries of the theory. As will now be explained, this somewhat complicated space can be usefully formulated in terms of the supermoduli space of super Riemann structures. The symmetries of the theory are the group of general coordinate transformations of the supermanifold and a group of transformations of the vielbein defined by Howe which are super analogues of the conformal transformations of a conventional Riemann metric. These transformations, which are referred to as super Weyl transformations, act on the vielbein in the following manner:

$$E_M{}^a \mapsto \Lambda E_M{}^a,$$
$$E_M{}^\alpha \mapsto \Lambda^{\frac{1}{2}} E_M{}^\alpha - \tfrac{i}{2}\Lambda^{-\frac{1}{2}}\gamma_a^{\alpha\beta}\mathcal{D}_\beta\Lambda. \tag{14.4}$$

Here the parameter Λ is an even G^∞ invertible scalar function on the supermanifold $\mathcal{M}$. These transformations, which were first derived by Howe, are slightly more complicated than a simple rescaling of the vielbein, but are the simplest which preserve the torsion constraints while placing no restriction on the parameter Λ. Thus one finds that one must integrate over the space of super Weyl and supersmooth equivalence classes on $\mathcal{M}$, $\mathcal{M}$ itself being determined by the choice of genus and spin structure on the underlying manifold $\mathcal{M}_{[\emptyset]}$. The link between the real $(2,2)$-dimensional geometry and the complex $(1,1)$-dimensional super Riemann surface concept of Baranov and Schwarz [6, 7] and Friedan [56] is provided by the important observation of Howe [76] that a local coordinate system can always be found in which the vielbein are super Weyl flat. A vielbein $E_M{}^A$ is said to be *super Weyl flat* if it is obtainable from the flat vielbein

$$\bar{E}_m{}^a = \delta_m^a \qquad \bar{E}_m{}^\alpha = 0$$
$$\bar{E}_\mu{}^a = 0 \qquad \bar{E}_\mu{}^\alpha = \delta_\mu^\alpha \tag{14.5}$$

by a super Weyl transformation. That is, $E_M{}^A$ is flat if it is of the form

$$E_M{}^a = \Lambda \bar{E}_M{}^a$$

$$E_M{}^\alpha = \Lambda^{\frac{1}{2}} \bar{E}_M{}^\alpha - i \bar{E}_M{}^a \gamma_a^{\alpha\beta} \bar{\mathcal{D}}_\beta \Lambda \tag{14.6}$$

for some parameter Λ.

14.2   The definition of a super Riemann surface

A super Riemann surface is defined to be a $(1,1)$-dimensional complex supermanifold which satisfies an additional superconformal condition.

Definition 14.2.1

(a) A *super Riemann structure* on a $(1,1)$-dimensional complex superanalytic supermanifold $\mathcal{M}$ is an atlas of superanalytic charts $\{(U_\alpha, \psi_\alpha) \,|\, \alpha \in \Upsilon\}$ on $\mathcal{M}$ such that whenever $U_\alpha \cap U_{\alpha'}$ is non-empty the corresponding coordinate transition function $(z_\alpha, \zeta_\alpha) \mapsto (z_{\alpha'}, \zeta_{\alpha'})$ is superconformal in the sense that the super derivative $\mathcal{D} = \frac{\partial}{\partial \zeta} + \zeta \frac{\partial}{\partial z}$ transforms covariantly with

$$\mathcal{D}_\alpha = \mathcal{D}_\alpha \zeta_{\alpha'} \, \mathcal{D}_{\alpha'} \,. \tag{14.7}$$

(b) A super Riemann surface is a $(1,1)$-dimensional complex superanalytic supermanifold $\mathcal{M}$ together with a super Riemann structure on $\mathcal{M}$. The coordinates in the preferred atlas are known as superconformal coordinates.

(c) A superconformal isomorphism of super Riemann surfaces $\mathcal{M}$ and $\mathcal{N}$ is a bijective mapping $f : \mathcal{M} \to \mathcal{N}$ which is a superanalytic diffeomorphism and which maps superconformal coordinates to superconformal coordinates.

The relationship between superconformal classes of super Riemann surfaces and super Weyl equivalence classes of $(2,2)$-dimensional real supermanifolds built on spin bundles then emerges in the following way. Starting with the real $(2,2)$-dimensional supermanifold $\mathcal{M}$, using Howe's result one chooses everywhere coordinates $(x^1, x^2; \theta^1, \theta^2)$ such that the vielbein $E_M{}^A$ are super Weyl flat. Combining these coordinates into complex coordinates $(z; \zeta)$ with $z = x^1 + ix^2$ and $\zeta, \bar{\zeta}$ chiral, one finds that changes of coordinate which preserve the super Weyl flatness of the vielbein are exactly the superconformal changes of coordinate [164, 61], so that a super Riemann structure has been obtained.

It is possible to obtain an explicit expression for a superconformal change of coordinate $(z; \zeta) \mapsto (\tilde{z}; \tilde{\zeta})$: taking the general superanalytic change of coordinate

$$\tilde{z} = f(z) + \zeta\beta(z), \qquad \tilde{\zeta} = \psi(z) + \zeta g(z) \tag{14.8}$$

and applying the superconformal condition (14.7), which implies that $\mathcal{D}_\alpha z_{\alpha'} = \zeta_{\alpha'}\mathcal{D}_\alpha\zeta_{\alpha'}$, shows that the functions $\beta(z)$ and $g(z)$ are determined by $f(z)$ and $\psi(z)$, explicitly

$$\tilde{z} = f(z) + \zeta\psi(z)\sqrt{f'(z)}, \qquad \tilde{\zeta} = \psi(z) + \zeta\sqrt{f'(z) + \psi(z)\psi'(z)}. \tag{14.9}$$

This shows that the underlying manifold of a super Riemann surface is a classical Riemann surface with coordinate transition functions $f_{[\emptyset]}$. It also shows that the supermanifold $S(M, \sqrt{K})$ constructed from a classical Riemann surface M, where K denotes the canonical line bundle of M and $\sqrt{K}$ denotes the choice of spin-bundle corresponding to a choice of square root of K, using the construction Theorem 8.1.1, is a super Riemann surface. In this case the even part f of the coordinate transition function is simply the Grassmann analytic continuation of the corresponding transition function on the Riemann surface, the sign of the square root $\sqrt{f'(z)}$ is determined by the chosen spin structure, and the odd part ψ is zero. Such a super Riemann surface will be referred to as the canonical super Riemann surface over M corresponding to the given spin structure $\sqrt{K}$. Not all super Riemann surfaces are canonical; in fact non-split super Riemann exist, as will emerge below, and the supermoduli space of super Riemann structures has both even and odd coordinates.

Three important examples of super Riemann surfaces, which play a key role in the uniformisation theory developed by Crane and Rabin [37], will now be described.

Example 14.2.2

(a) The complex superspace $\mathbb{C}_S^{1,1}$ with natural coordinates (z, ζ) is a super Riemann surface, which will be denoted SC.

(b) The canonical Super Riemann surface corresponding to the unique spin structure on the Riemann sphere C^* will be denoted SC^*. Corresponding to the standard coordinates $z, \tilde{z}$ on C^*, with $z = \frac{1}{\tilde{z}}$, there are coordinates (z, ζ) and $(\tilde{z}, \tilde{\zeta})$ with

$$z = \frac{1}{\tilde{z}} \quad \text{and} \quad \zeta = i\frac{1}{\tilde{z}}\tilde{\zeta}. \tag{14.10}$$

(c) The canonical Super Riemann surface corresponding to the unique spin structure on the upper half plane U will be denoted SU.

These three super Riemann surfaces are the only cases whose body is simply connected.

The simplest example of a non-split super Riemann surface is the super torus , which uses to the odd spin structure on a classical torus.

Definition 14.2.3   The super torus $\mathcal{T}(a, \delta)$ is the super Riemann surface $\mathbb{C}_S^{1,1}/\Gamma$ where Γ is the 'super lattice' action

$$(z; \zeta) \rightarrow (z + n + m(a + \zeta\delta); \zeta + m\delta). \tag{14.11}$$

Here m and n are integers, a is an even complex Grassmann parameter with $\Im(a_{[\emptyset]}) > 0$ and δ is an odd parameter.

14.3   The supermoduli space of super Riemann surfaces

This section relies very heavily on Crane and Rabin's pioneering work on super Riemann surfaces [37]. The first step in the investigation of supermoduli space is the uniformisation theory of super Riemann surfaces, of which the key result is Theorem 14.3.2. This depends on first identifying the simply connected super Riemann surfaces, and then identifying the appropriate super analogue of the Möbius transformations of the complex plane.

In the proof of the following proposition, which establishes that SC, SC^* and SU (Example 14.2.2) are the only super Riemann surfaces with simply connected body, use is made of the concrete supermanifold formalism, working level by level in the Grassmann algebra $\mathbb{C}_S$.

Proposition 14.3.1   *Suppose that $\mathcal{M}$ is a simply connected super Riemann surface. Then $\mathcal{M}$ is superconformally diffeomorphic to SC, SC^* or SU.*

Proof   Let $\{U_\alpha | \alpha \in \Upsilon\}$ be an open cover of $\mathcal{M}$ by coordinate neighbourhoods, with local coordinates on each U_α denoted (z_α, ζ_α). Suppose that α, β and γ in Υ are such that $U_\alpha \cap U_\beta \cap U_\gamma \neq \emptyset$, and that functions $f^{\alpha\beta}, \psi^{\alpha\beta}, f^{\beta\gamma}, \psi^{\beta\gamma}$ and $f^{\alpha\gamma}, \psi^{\alpha\gamma}$ determine the changes of coordinate $(z_\beta, \zeta_\beta) \to (z_\alpha, \zeta_\alpha)$, $(z_\gamma, \zeta_\gamma) \to (z_\beta, \zeta_\beta)$ and $(z_\gamma, \zeta_\gamma) \to (z_\alpha, \zeta_\alpha)$ respectively as in (14.9). Then consistency of changing from the α system to the γ system either directly or via the β system gives the cocycle condition

$$f^{\alpha\gamma} = (f^{\alpha\beta} \circ f^{\beta\gamma}) + \psi^{\beta\gamma}(\psi^{\alpha\beta} \circ f^{\beta\gamma})\sqrt{(f^{\alpha\beta\prime} \circ f^{\beta\gamma})}$$

$$\psi^{\alpha\gamma} = (\psi^{\alpha\beta} \circ f^{\beta\gamma}) + \psi^{\beta\gamma}\sqrt{(f^{\alpha\beta\prime} \circ f^{\beta\gamma}) + (\psi^{\alpha\beta} \circ f^{\beta\gamma})(\psi^{\alpha\beta\prime} \circ f^{\beta\gamma})}.$$

$$(14.12)$$

Analysing this level by level in $\mathbb{C}_S$, level zero gives

$$f^{\alpha\gamma}_{[\emptyset]} = f^{\alpha\beta}_{[\emptyset]} \circ f^{\beta\gamma}_{[\emptyset]} \tag{14.13}$$

which simply confirms that $f^{\alpha\gamma}_{[\emptyset]}, f^{\alpha\beta}_{[\emptyset]}$ and $f^{\beta\gamma}_{[\emptyset]}$ are the coordinate transition functions of the body of $\mathcal{M}$. At level one the coefficient of $\beta_{[i]}$ gives

$$\psi^{\alpha\gamma}_i = \psi^{\alpha\beta}_i \circ f^{\beta\gamma}_{[\emptyset]} + \psi^{\beta\gamma}_i \sqrt{f^{\alpha\beta}_{[\emptyset]}{}' \circ f^{\beta\gamma}_{[\emptyset]}}. \tag{14.14}$$

Multiplying through by the local representative e^α of a section e of the spinor bundle on $\mathcal{M}_{[\emptyset]}$ gives

$$e^\alpha \psi^{\alpha\gamma}_i = e^\alpha \psi^{\alpha\beta}_i \circ f^{\beta\gamma}_{[\emptyset]} + e^\beta \psi^{\beta\gamma}_i, \tag{14.15}$$

where the fact that $e^\beta = e^\alpha \sqrt{f^{\alpha\beta}_{[\emptyset]}{}' \circ f^{\beta\gamma}_{[\emptyset]}}$ has been used. This shows that $\{e^\alpha \psi^{\alpha\gamma}_i | (\alpha, \gamma) \in \Upsilon \times \Upsilon\}$ are the local representatives of a cocycle in Čech cohomology group of $\mathcal{M}_{[\emptyset]}$ with coefficients in the sheaf of cross sections of the spinor bundle of $\mathcal{M}_{[\emptyset]}$. Crane and Rabin observe that this cohomology is trivial, and hence that there exist $\{\eta^\alpha | \alpha \in \Upsilon\}$ such that

$$e^\alpha \psi^{\alpha\gamma}_i = e^\gamma \eta^\gamma - e^\alpha \eta^\alpha. \tag{14.16}$$

This allows a superconformal redefinition of coordinates $(\tilde{z}^\alpha, \tilde{\zeta}_\alpha)$ with $f(z_\alpha) = z_\alpha$ and $\psi(z) = \eta^\alpha(z_\alpha)\beta_{[i]}$ which leads to $\tilde{\psi}^{\alpha\gamma}_i = 0$. Similar arguments allow successive conformal redefinition of coordinates in such a way that all soul terms in the redefined $f^{\alpha\beta}$, and all terms in the redefined $\psi^{\alpha\beta}$, become zero. This establishes the result. ∎

The group that plays the role of the group of Möbius transformations in the super setting is the group $\mathrm{SPL}(\mathbb{C}, 2)$ of superconformal transformations of SC of the form (14.9) with

$$f(z) = \frac{az+b}{cz+d}, \quad \text{and} \quad \psi(z) = \frac{\gamma z + \delta}{cz+d} \qquad (14.17)$$

where a, b, c and d are even elements of $\mathbb{R}_S$ and satisfy $ad - bc = 1$ while γ and δ are odd. This group may be obtained by exponentiation of the superconformal algebra with even generators L_{-1}, L_1, L_0 and odd generators $G_{\frac{1}{2}}, G_{-\frac{1}{2}}$ and bracket

$$[L_m, L_n] = (n - m)L_{n+m}$$

$$[L_m, G_r] = \left(\frac{m}{2} - r\right) G_{m+r}$$

$$[G_r, G_s] = 2L_{r+s} . \qquad (14.18)$$

As first shown by Crane and Rabin [37], this is the group of superconformal automorphisms of the super Riemann sphere. The proof of the uniformisation theorem is not an entirely straightforward generalisation of the classical one because there are superconformal automorphisms of SC and SU which are not in $\mathrm{SPL}(\mathbb{C}, 2)$. However Hodgkin [73] has shown that all elements in these automorphism groups can be obtained by conjugation of elements of $\mathrm{SPL}(\mathbb{C}, 2)$, which irons out this difficulty, and allows the following uniformisation theorem.

Theorem 14.3.2 *Any super Riemann surface is either SC^* or the quotient of either SC or SU by a discrete subgroup Γ of the group $\mathrm{SPL}(\mathbb{C}, 2)$ acting properly and discontinuously with respect to the DeWitt topology.*

Outline of proof If $\mathcal{M}$ is a super Riemann surface with body $\mathcal{M}_{[\emptyset]}$, then its universal cover $\widetilde{\mathcal{M}}$ is a super Riemann surface whose body is $\widetilde{\mathcal{M}_{[\emptyset]}}$. Also, $\widetilde{\mathcal{M}}$ may be given a superconformal structure in such a way that the covering group of $\mathcal{M}$ acts by superconformal transformations. Thus any super Riemann surface will have universal cover either SC^*, SC or SU, and will be the quotient of one of these super Riemann surfaces by a discrete subgroup Γ of the group of superconformal automorphisms of its universal cover which acts properly and discontinuously with respect to the DeWitt topology.

Since no subgroup of the Möbius group acts properly discontinuously on C^*, there cannot be any subgroup of $\mathrm{SPL}(\mathbb{C}, 2)$ which acts properly discontinuously on SC^* with respect to the DeWitt topology. Thus, since $\mathrm{SPL}(\mathbb{C}, 2)$ is the group of superconformal automorphisms of SC^*, the only super Riemann surface with universal cover SC^* is SC^* itself. Use of

Hodgkin's result [73] that up to conjugation the group of automorphisms of SC and SU are both $\mathrm{SPL}(\mathbb{C}, 2)$ establishes the theorem. ∎

Thus, using the fact that the fundamental group of a Riemann surface of genus g has $2g$ generators and one relation, it can be deduced that super Teichmüller space has real dimension $(6g - 6, 4g - 4)$. (There will be points in Teichmüller space where the even moduli are zero, but the odd moduli are non-zero, for which the corresponding super Riemann surface will be singular in the sense that it will have to be formulated using the fine topology as in Example 5.3.2.) The structure of supermoduli space has received some attention [98, 75, 45] but is not fully resolved. The structure of supermoduli space is important because the functional integral approach to the quantization of spinning strings is expected to lead ultimately to an integral over this space; as has been seen in Chapter 11 there may be difficulty in defining such integrals if the supermoduli space is not split.

14.4 Contour integration on super Riemann surfaces

In this section a theory of contour integration for superconformal curves in SC is developed. The superconformal structure of a super Riemann surface then allows an object known as half-form whose contour integration possesses many of those features of classical integration which can be difficult to preserve on a supermanifold. The integral is defined using the double integral with odd and even limits developed in Section 11.6.

Starting with SC (which is simply $\mathbb{C}_S^{1,1}$ equipped with a canonical super Riemann surface structure), the trivial super Riemann surface, a superconformal curve is defined to be a mapping

$$
\begin{aligned}
C : U &\to \mathbb{C}_S^{1,1} \\
(t; \tau) &\mapsto (c(t; \tau); \gamma(t; \tau))
\end{aligned}
\tag{14.19}
$$

(where U is open in $\mathbb{R}_S^{1,1}$) that is real superconformal so that (extending (11.88) by allowing a complex codomain) there exist C^∞ functions $f : \mathbb{R} \to \mathbb{C}_{S1}$ and $\phi : \mathbb{R} \to \mathbb{C}_{S0}$ with

$$
C(t; \tau) = \left(f(t) + \tau \phi(t) \sqrt{f'(t)}, \tau \phi(t) + \sqrt{f'(t) + \phi(t) \phi'(t)} \right).
\tag{14.20}
$$

Suppose that $C(a; \alpha) = (p; \pi)$ and $C(b; \beta) = (r; \rho)$. The integral of a function $f : \mathbb{C}_S^{1,1} \to \mathbb{C}_S$ from $(p; \pi)$ to $(r; \rho)$ along C is defined to be

$$\int_{C[p;\pi;r;\rho]} f(z; \zeta) \mathcal{D}Z = \int_\alpha^\beta \int_a^b f(C(t; \tau)) \mathcal{D}_T \gamma(t; \tau) \mathcal{D}T. \qquad (14.21)$$

This integral has two useful transformation properties. First, it is invariant under superconformal reparameterisation of the curve C. That is, if K is an invertible real superconformal map of $\mathbb{R}_S^{1,1}$ onto $\mathbb{R}_S^{1,1}$, then

$$\int_{C[p;\pi:r;\rho]} f(z; \zeta) \mathcal{D}Z = \int_{C \circ K[p;\pi:r;\rho]} f(z; \zeta) \mathcal{D}Z. \qquad (14.22)$$

This result follows from the change of variable rule (11.89) and the chain rule (11.90) for superconformal functions.

Additionally further use of the change of variable rule shows that this contour integral transforms naturally under superconformal change of coordinate on SC; explicitly suppose that $H : SC \to SC$ is superconformal with $H(z; \zeta) = (h(z; \zeta); \eta(z; \zeta))$. Then, if G is a supersmooth function on SC,

$$\int_{H \circ C[H(p;\pi):H(r;\rho)]} G(Z') \mathcal{D}Z' = \int_{C[p;\pi:r;\rho]} G(H(Z)) \mathcal{D}_Z \eta \mathcal{D}Z. \qquad (14.23)$$

This rule follows from the chain rule (11.90) which is also valid when $K : SC \to SC$ and F has domain SC.

The contour integral of a superanalytic function obeys a Cauchy theorem.

Theorem 14.4.1 *Suppose that C is a closed superconformal curve in SC such that the closed curve $C_{[0]}$ is a Jordan curve in $\mathbb{C}$. Let U be an open set in $\mathbb{C}$ which contains $C_{[0]}$ and its interior and F be a superanalytic function on $(\epsilon_{1,1})^{-1}(U)$. Then*

$$\int_C F(Z) \mathcal{D}Z = 0. \qquad (14.24)$$

The theorem may be proved by expanding in powers of the odd parameter τ of the curve and applying the classical Cauchy theorem.

It is possible to define a winding number for a closed superconformal curve, and also to use the Cauchy theorem to provide an integral representation of analytic functions. The winding number of the superconformal curve C about the point $(r; \rho)$ in SC is defined to be

$$N(C) = \frac{1}{2\pi i} \int_C \frac{\zeta - \rho}{z - r} \mathcal{D}Z .$$
(14.25)

Application of the Cauchy theorem then shows that if F satisfies the conditions of Theorem 14.4.1

$$F(r; \rho) = \frac{2\pi i}{N(C)} \int_C F(z; \zeta) \mathcal{D}Z .$$
(14.26)

The transformation rule (14.23) makes it clear that the type of object which can be consistently integrated along a superconformal curve must have a local representative which transforms according to the superconformal scale factor $\mathcal{D}\zeta$. Such an object will now be defined.

Definition 14.4.2   A *half form* W on a super Riemann surface M is an object which in a local coordinate system has the expression $W(Z)\mathcal{D}Z$ and the functions $W(Z)$, $W(Z')$ on overlapping coordinate patches are related by

$$W(Z) = W(Z')\mathcal{D}_Z\zeta'$$
(14.27)

where $Z'(z; \zeta) = (z'(Z); \zeta'(Z))$.

A definition of half form in terms of the reduction of the bundle of superanalytic frames of $\mathcal{M}$ which exists by virtue of the superconformal structure is also possible. The transformation property (14.27) of half forms together with the change of variable rule (14.23) makes it immediately clear that a half form has a well defined integral along a superconformal curve in $\mathcal{M}$.

14.5   Fields on super Riemann surfaces

Many of the applications of the theory of super Riemann surfaces to the calculation of multiloop contributions in string theory depend on certain spaces of fields on super Riemann surfaces having a super vector space structure that they do not in fact possess. Not only do these spaces lack this structure, but also the nature of these spaces may vary as one moves around the moduli space of super Riemann surfaces corresponding to fixed genus and spin structure on the underlying Riemann surface. This difficulty, which undermines some of the methods used, was first observed in the literature by Giddings and Nelson [62] and then extensively analysed by Hodgkin [74].

On a split super Riemann surface the problem does not occur. Denote by $\check{M}$ the canonical super Riemann surface $S(M, \sqrt{K})$ with compact body M and spin structure $\sqrt{K}$, and suppose that $(z_\alpha; \zeta_\alpha)$ are local coordinates on $\check{M}$. Then the local representative of an even superholomorphic function g on $\check{M}$ takes the form

$$g_\alpha(z_\alpha; \zeta_\alpha) = p + \zeta_\alpha \pi_\alpha(z_\alpha) \qquad (14.28)$$

where p is a constant element of $\mathbb{C}_{S0}$ and π_α is the local representative of a $\mathbb{C}_{S1}$-valued section of the spin bundle $\sqrt{K}$. As a result the space of superholomorphic functions has the structure

$$\mathcal{O}(\check{M}) \cong (\mathbb{C}_{S0} \otimes \mathbb{C}) \oplus (\mathbb{C}_{S1} \otimes \Gamma(\sqrt{K})) . \qquad (14.29)$$

Spaces of cross-sections of other bundles over $S(M, \sqrt{K})$ have analogous super vector space structure.

However, although such a structure appears to be required for the superholomorphic quantization of the spinning string, the space of superholomorphic functions on an arbitrary super Riemann surface does not in general have the required structure. A simple example where this structure breaks down is on the super torus of example Example 14.2.3 [62, 74]. In this case, as is clear from (14.11), the most general even superholomorphic function takes the form

$$g(z; \zeta) = c + \gamma\zeta \qquad (14.30)$$

where γ is a constant element of $\mathbb{C}_{S1}$ which must satisfy $\gamma\delta = 0$. This condition prevents the space of functions having the simple structure of a $(1,1)$-dimensional super vector space except in the split space where $\delta = 0$.

In [129] a modified definition of *superconformal function* was given leading to a space which does have the required structure. The rough idea is that a superconformal function on a super Riemann surface is a cross-section of a $(1,1)$ dimensional holomorphic vector bundle on the surface which has local representatives of the form $\begin{pmatrix} g_\alpha \\ \mathcal{D}_\alpha g_\alpha \end{pmatrix}$. (Baranov and Schwarz emphasise the importance of the pair $(g, \mathcal{D}g)$ in their work on poles and zeroes on super Riemann surfaces [6].) The required vector bundle, which was first constructed by D'Hoker and Phong [44], is constructed in the following proposition.

Proposition 14.5.1 *Suppose that $\mathcal{M}$ is a super Riemann surface with superconformal structure $\{(U_\alpha, \psi_\alpha)|\alpha \in \Upsilon\}$. For α, β in Υ with $U_\alpha \cap U_\beta \neq \emptyset$*

define $M_{\alpha\beta}(z_\beta; \zeta_\beta) = \mathcal{D}_\beta \zeta_\alpha(z_\beta; \zeta_\beta)$ so that $\mathcal{D}_\beta = M_{\alpha\beta}(z_\beta; \zeta_\beta)\mathcal{D}_\alpha$. Also, with the notation of Proposition 14.3.1, define the $\mathrm{GL}(1,1;\mathbb{C})$-valued function on $\psi_\beta(U_\alpha \cap U_\beta)$ by

$$\mathcal{R}_{\alpha\beta}(z_\beta; \zeta_\beta) = \begin{pmatrix} 1 & (M_{\alpha\beta}(z_\beta))^{-1}\psi_{\alpha\beta} \\ 0 & (M_{\alpha\beta}(z_\beta))^{-1} \end{pmatrix}. \tag{14.31}$$

Then these functions are the transition functions of a $(1,1)$-dimensional super vector bundle on $\mathcal{M}$. This super vector bundle will be denoted $E_\mathcal{M}$.

Outline of proof Using (14.12), it may be shown that, for each α in Υ, $\mathcal{R}_{\alpha\alpha}(z_\alpha; \zeta_\alpha)$ is the identity matrix. Also, for each α, β in Υ such that $U_\alpha \cap U_\beta \neq \emptyset$, $\mathcal{R}_{\alpha\beta}(z_\beta; \zeta_\beta)^{-1} = \mathcal{R}_{\beta\alpha}(z_\alpha; \zeta_\alpha)$ and for each $\alpha, \beta\gamma$ in Υ such that $U_\alpha \cap U_\beta \cap U_\gamma \neq \emptyset$,

$$\mathcal{R}_{\alpha\gamma}(z_\gamma; \zeta_\gamma) = \mathcal{R}_{\alpha\beta}(\psi_{\beta\gamma}(z_\alpha, \zeta_\beta))\mathcal{R}_{\beta\gamma}(z_\gamma; \zeta_\gamma) \tag{14.32}$$

∎

The notion of superconformal function can now be defined.

Definition 14.5.2 A *superconformal* function is a cross-section of the super vector bundle $E_\mathcal{M}$ such that at every point there exists a local coordinate system (z_α, ζ_α) in which the components $\begin{pmatrix} g_\alpha(z_\alpha, \zeta_\alpha) \\ \gamma_\alpha(z_\alpha, \zeta_\alpha) \end{pmatrix}$ of the local representative satisfy $\gamma_\alpha(z_\alpha, \zeta_\alpha) = \mathcal{D}_\alpha g_\alpha(z_\alpha, \zeta_\alpha)$. The space of such fields is denoted $\mathrm{SC}(\mathcal{M})$.

The point of this construction is that it leads to an object whose first order term in the Taylor expansion in ζ has sufficient freedom to allow the desired super vector space structure for $\mathrm{SC}(\mathcal{M})$, so that the following theorem can be established.

Theorem 14.5.3

$$\mathrm{SC}(\mathcal{M}) \cong (\mathbb{C}_{S0} \otimes \mathbb{C}) \oplus (\mathbb{C}_{S1} \otimes \Gamma(\sqrt{K})). \tag{14.33}$$

An elegant proof of this result by Rabin, much improving on that of [129], may be found in [114].

Chapter 15

Path integrals on supermanifolds

Historically, one of the first uses of anticommuting variables in physics was by Martin [103, 102], who extended Feynman's 'sum over histories' approach to the quantization of bosonic particles to fermions. These methods have been widely developed, leading to a battery of techniques described in a number of books including those of Roepstorff[117], Schulman[137] and Swanson[148]. Inoue and Maeda have considered fermionic path integrals in the contect of their work on superanalysis [82].

In this chapter a brief review is given of these ideas together with their more formal expression in terms of analogues of Brownian motion and Wiener measure on a purely odd superspace. This is the content of Sections 15.1 and 15.2. In Section 15.3 these constructions are then combined with classical Brownian motion and Wiener measure to develop path integrals on superspace. Stochastic calculus is extended to superspace in Section 15.4, and these methods used in Section 15.5 to define Brownian paths on supermanifolds which can be applied to various geometric operators. With the aim of ensuring that the material is accessible to readers without extensive expertise in probability theory, a rather more informal style, without full analytic rigour, is adopted in places.

The generalised probabilistic methods described here can be applied to systems with supersymmetry, BRST-quantized systems such as those in Chapter 16 as well as to the study of some classical geometric operators as in Chapter 17.

15.1 Path integrals and fermions

Fermionic degrees of freedom do not appear in classical physics, they are purely quantum phenomena. However, in techniques for quantum theories which require a classical system which is then quantized, 'classical' fermions are required; one way of constructing such objects is to use anticommuting variables, leading to canonical anticommutation relations on quantization. These variables do not directly model physical reality, and the models involved are sometimes referred to as pseudoclassical.

As an example of such a system, a fermionic oscillator will be considered. This is a system whose phase space is $\mathbb{R}_S^{0,2}$, with anticommuting phase space variables ψ^1, ψ^2 and Poisson bracket

$$\left\{\psi^i, \psi^j\right\} = \delta^{ij} . \tag{15.1}$$

The Hamiltonian for the system is

$$H = ia\psi^1\psi^2 \tag{15.2}$$

where a is a constant. The system is quantized by taking states to be functions on $\mathbb{R}_S^{0,2}$; taking θ^1, θ^2 as natural coordinates, the fermionic operators $\psi^i : i = 1, 2$ are represented as differential operators:

$$\psi^i = \tfrac{1}{\sqrt{2}} \left(\theta^i + \frac{\partial}{\partial\theta^i} \right) . \tag{15.3}$$

Direct calculations show that these operators obey the canonical anticommutation rule

$$\left[\psi^i, \psi^j\right] = \delta^{ij} \tag{15.4}$$

which is the expected quantization of the Poisson bracket (15.1).

The Hamiltonian operator is thus the second order differential operator

$$H = \tfrac{1}{2}ia \left(\theta^1\theta^2 + \left(\theta^1\frac{\partial}{\partial\theta^2} - \theta^2\frac{\partial}{\partial\theta^1} \right) + \frac{\partial}{\partial\theta^1}\frac{\partial}{\partial\theta^2} \right) \tag{15.5}$$

and the system is solved if one can find an expression for the evolution operator $\exp(-Ht)$. In this case it is easily seen by direct calculation (using (15.4) to show that $\left(\psi^1\psi^2\right)^2 = 1$) that

$$\exp(-Ht) = \cosh at + \psi^1\psi^2 \sinh at . \tag{15.6}$$

This system has the usual feature of fermionic mechanics that the Hamiltonian has no kinetic term. The free Hamiltonian, corresponding to $a = 0$, is simply zero.

15.2 Fermionic Brownian motion

There are various ways in which one can define classical Brownian motion
and Wiener measure; an approach which allows a fermionic analogue is
to define it in terms of finite-dimensional marginal distributions. In this
approach, using the notation $\{b_t | t > 0\}$ for Brownian motion, the joint
distribution of $(b_{t_1}, \ldots, b_{t_N}), 0 < t_1 < \cdots < t_N$ is

$$P_{s_1}(0, x_1) P_{s_2}(x_1, x_2) \ldots P_{s_N}(x_{N-1}, x_N) \, \mathrm{d}^N x \qquad (15.7)$$

where $s_1 = t_1, \, s_2 = t_2 - t_1, \ldots, s_N = s_N - s_{N-1}$ and

$$P_t(x, y) = \frac{1}{\sqrt{2\pi t}} \exp\left(\frac{-(x - y)^2}{2t}\right). \qquad (15.8)$$

Here $x = (x^1, \ldots, x^m)$ and $y = (y^1, \ldots, y^m)$ are both points in $\mathbb{R}^m$,
$(x - y)^2 = \sum_{i=1}^{m}(x^i - y^i)(x^i - y^i)$, and the distributions are those for
Brownian motion in m dimensions. This definition means that if F is a
function on the space of paths in $\mathbb{R}^m$ which only depends on the path at a
finite set of times $t_1, \ldots, t_N$, (so that $F(b_t) = F(b_{t_1}, b_{t_2}, \ldots, b_{t_N})$) and $\mathrm{d}\mu_b$
denotes Wiener measure, then

$$\int \mathrm{d}\mu_b \, F$$
$$= \int \mathrm{d}^m x_1 \ldots \mathrm{d}^m x_N \, P_{s_1}(0, x_1) P_{s_2}(x_1, x_2) \ldots P_{s_N}(x_{N-1}, x_N)$$
$$\times \quad F(x_1, x_2, \ldots, x_N).$$
$$(15.9)$$

A useful example, which will be important in Section 15.4, is that

$$\int \mathrm{d}\mu_b \, b_s b_t = \min(s, t). \qquad (15.10)$$

Comparing (15.9) to the usual heuristic discrete time expression of the path
integral shows that Wiener measure corresponds to the kinetic term in the
action. The function $P_t(x, y)$ is the kernel of the evolution operator for
the free particle. The fermionic analogue of this construction will now be
described. It is built in a similar manner from the evolution operator of
the free fermion, which is simply the identity operator. The function which
plays the role of $P_t(x, y)$ is thus the kernel of the identity operator; however, for reasons which will become clear below, it is defined in phase space
rather than simply position space; put another way, the Fourier transform

variables are retained as part of the measure. Thus fermionic Wiener measure in n dimensions is built from the functions $\pi_t : \mathbb{R}_S^{0,3n} \to \mathbb{R}_S, t > 0$ with

$$\pi_t\,(\theta, \phi, \rho) = \exp\left(-i\rho \cdot (\theta - \phi)\right) \qquad (15.11)$$

where $\rho \cdot (\theta - \phi) = \sum_{i=1}^n \rho_i(\theta^i - \phi^i)$. In this chapter, to avoid complicated factors, it is assumed that n is even. As a result

$$\int \mathrm{d}^n \rho \exp\left(-i\rho \cdot (\theta - \phi)\right) = \delta(\theta - \phi) \qquad (15.12)$$

so that integration of π_t with respect to the Fourier transform or momentum variable which is its third argument gives the delta function which is the kernel of the identity operator as required by analogy with bosonic Wiener measure. Fermionic Brownian motion in n dimensions is then defined to be the $\mathbb{R}_S^{0,2n}$-valued process with finite-dimensional joint distributions having the form

$$\Phi_{t_1,\ldots,t_N}(\theta_1, \rho_1, \ldots, \theta_N, \rho_N)$$
$$= \pi_{s_1}(0, \theta_1, \rho_1)\,\pi_{s_2}(\theta_1, \theta_2, \rho_2) \ldots \pi_{s_N}(\theta_{N-1}, \theta_N, \rho_N) \qquad (15.13)$$

where as before $s_1 = t_1$, $s_2 = t_2 - t_1, \ldots, s_N = s_N - s_{N-1}$. Of course this does not define a true measure, however a notion of Grassmann probability measure can be defined provided that the distribution functions satisfy certain consistency conditions, specifically

$$\int \mathrm{d}^n \theta_1 \ldots \mathrm{d}^n \theta_N \mathrm{d}^n \rho_1 \ldots \mathrm{d}^n \rho_N \Phi_{t_1,\ldots,t_N}(\theta_1, \rho_1, \ldots, \theta_N, \rho_N) = 1$$

and

$$\int \mathrm{d}^n \theta_r \mathrm{d}^n \rho_r \Phi_{t_1,\ldots,t_N}(\theta_1, \rho_1, \ldots, \theta_N, \rho_N) =$$
$$\Phi_{t_1,\ldots,t_{r-1},t_{r+1},\ldots,t_N}(\theta_1, \rho_1, \ldots, \theta_{r-1}, \rho_{r-1}, \theta_{r+1}, \rho_{r+1}, \ldots, \theta_N, \rho_N)\,.$$
$$(15.14)$$

The corresponding process is denoted (θ_t, ρ_t). A limiting process then allows a sufficiently general notion of Grassmann random variable to be defined. Further details may be found in [123, 127]. In particular, given a function V on $\mathbb{R}_S^{0,2n}$ the random variable $\exp\left(-\int_o^t V(\theta_s, \rho_s)\mathrm{d}s\right)$ can be defined in the expected way.

This measure can be used to give a fermionic Feynman-Kac formula for a Hamiltonian of the form $H = \sum_{\underline{\mu} \in M_n} a_{\underline{\mu}}\, \psi^{\underline{\mu}}$. A limiting process quite close to the physicist's heuristic argument shows that [123, 127]

$$\exp\left(-Ht\right) f(\xi) = \int \mathrm{d}\mu_f \exp\left(-\int_o^t V(\xi + \theta_s + \rho_s)\mathrm{d}s\right) f(\xi + \theta_t). \quad (15.15)$$

The reason for this result is that the path integral on the right hand side is (broadly speaking) the limit as $N \to \infty$ of the discrete time expression

$$\int \prod_{r=1}^N \left(\mathrm{d}^n\theta_r \mathrm{d}^n\rho_r \ \pi_{s_r}\left(\theta_{r-1}, \theta_r, \rho_r\right)\right)$$

$$\times \ \exp\left(-\sum_{r=1}^N V(\xi + \theta_r + \rho_r)s_r\right) f(\xi + \theta_N)$$

$$=$$

$$\int \prod_{r=1}^N \left(\mathrm{d}^n\theta_r \mathrm{d}^n\rho_r\right)$$

$$\times \ \exp\left(-\sum_{r=1}^N \left(\rho_r \cdot \frac{\theta_r - \theta_{r-1}}{s_r} + V(\xi + \theta_r + \rho_r)\right) s_r\right) f(\xi + \theta_N)$$

$$(15.16)$$

where, for $r = 0 \ldots N$, $t_r = \frac{rt}{N}$ and $s_r = t_r - t_{r-1}$ is as before.

It should be emphasised that the constructions in this section do not generalise all aspects of the classical probabilistic theory of Brownian motion; only those aspects which are expressed in terms of expectations, or equivalently as integrals with respect to Wiener measure, are included in this formalism. This is sufficient to extend to the fermionic setting stochastic methods for studying heat kernels (or, in quantum mechanics terminology, evolution operators) and diffusion equations.

15.3 Brownian motion in superspace

The fermionic Brownian motion described in the previous section can be combined with classical Brownian motion to give a notion of Brownian motion in (m, n)-dimensional superspace. This is achieved by taking the direct product of bosonic and fermionic Wiener measure in m and n dimensions respectively, obtaining a measure (in a generalised sense) on the space $\left(\mathbb{R}_S^{m,2n}\right)^{(0,\infty)}$ of paths in $\mathbb{R}_S^{m,2n}$. The corresponding super Wiener measure will be denoted $\mathrm{d}\mu_s$. Super Wiener measure has finite dimensional distributions

$$
\begin{aligned}
F_{t_1,\ldots,t_N} &(x_1, \theta_1, \rho_1, \ldots, x_N, \theta_N, \rho_N) \\
&= P_{s_1}(0, x_1)\, P_{s_2}(x_1, x_2) \ldots P_{s_N}(x_{N-1}, x_N) \\
&\quad \times \pi_{s_1}(0, \theta_1, \rho_1)\, \pi_{s_2}(\theta_1, \theta_2, \rho_2) \ldots \pi_{s_N}(\theta_{N-1}, \theta_N, \rho_N)\,. \quad (15.17)
\end{aligned}
$$

Because all integration over even variables is of rapidly decreasing functions, it is not necessary to dwell on any distinction between real (or complex) variables and even Grassmann variables.

Using this measure a Feynman-Kac formula for Hamiltonians of the form

$$
H = \tfrac{1}{2}p^2 + V(x, \psi) \qquad (15.18)
$$

can be constructed. (The potential V will need to satisfy certain conditions which will not be considered here.) The Feynman-Kac formula takes the form

$$
\begin{aligned}
\exp&(-Ht)\, f(x; \xi) \\
&= \int \mathrm{d}\mu_f \exp\left(-\int_o^t V(x + b_t; \xi + \theta_s + \rho_s)\mathrm{d}s\right) f(x + b_t; \xi + \theta_t)\,.
\end{aligned}
$$
$$(15.19)$$

Again this result can be roughly proved by observing that the path integral on the right hand side is (broadly speaking) the limit as $N \to \infty$ of the discrete time expression

$$
\int \prod_{r=1}^{N} (\mathrm{d}^n x_r \mathrm{d}^n \theta_r \mathrm{d}^n \rho_r\; P_{s_r}(x_{r-1}, x_r)\, \pi_{s_r}(\theta_{r-1}, \theta_r, \rho_r))
$$

$$
\times \exp\left(-\sum_{r=1}^{N} V(x + x_r, \xi + \theta_r + \rho_r)s_r\right) f(x + x_N; \xi + \theta_N)
$$

$$
= \int \prod_{r=1}^{N} (\mathrm{d}^n x_r \mathrm{d}^n \theta_r \mathrm{d}^n \rho_r)
$$

$$
\times \exp\left(-\sum_{r=1}^{N} \left(\frac{(x_r - x_{r-1})^2}{2s_r^2} + \rho_r \frac{\theta_r - \theta_{r-1}}{s_r} + V(x + x_r; \xi + \theta_r + \rho_r)\right) s_r\right)
$$

$$
\times f(x + x_N; \xi + \theta_N)\,. \qquad (15.20)
$$

In order to extend these techniques to supermanifolds, stochastic calculus must be developed to include fermionic Brownian motion. This is the subject of the next section.

15.4 Stochastic calculus in superspace

In the setting of purely classical or bosonic Brownian motion, stochastic calculus makes possible the extension of Feynman-Kac formulae to a considerably wider class of Hamiltonians, or, in mathematical language, to a larger class of second order elliptic partial differential equations. Important techniques from stochastic calculus for this purpose include the Itô integral and stochastic differential equations. In this section it will be shown that fermionic paths can be incorporated into stochastic calculus. Because fermionic Wiener measure includes the Fourier transform variable, it is not necessary to develop any analogue of Itô integral along fermionic Brownian paths, indeed these paths are far too irregular for any useful integral to be possible. An object which is required is the integral of an adapted stochastic process on the full super Wiener space along bosonic Brownian paths.

Because of the somewhat restricted version of probability theory being used, a simple definition of an adapted process is sufficient. A stochastic process on super Wiener space is a collection of random variables $\{A_t | t \in (0, \infty)\}$. Loosely speaking, a random variable on (m, n)-dimensional super Wiener space is a function on the space of paths $\left(\mathbb{R}_S^{m,2n} \right)^{(0,\infty)}$ which can be integrated with respect to super Wiener measure; the process is said to be *adapted* if for each t in $(0, \infty)$ the process A_t depends only on values of paths at times no greater than t. Adapted processes are important because they can be integrated both with respect to t and along bosonic Brownian paths, in a manner which will now be defined, again in the rather rough style of this chapter.

Definition 15.4.1 Suppose that $\{A_t | t \in (0, \infty)\}$ and, for $a = 1, \ldots, m$, $\{B_t^a | t \in (0, \infty)\}$ are adapted processes on (m, n)-dimensional super Wiener space. Then the *time integral* of A_t and the *Itô integral* of B_t are defined by

(a)

$$\int_o^t A_s \, \mathrm{d}s = \lim_{N \to \infty} \sum_{r=0}^{2^N - 1} \frac{t}{2^N} A_{\frac{rt}{2^N}} \qquad (15.21)$$

(b)

$$\int_o^t \sum_{a=1}^m B_s^a \, db_s^a = \lim_{N \to \infty} \sum_{r=0}^{2^N-1} \sum_{a=1}^m B_{\frac{rt}{2^N}}^a \left(b_{\frac{(r+1)t}{2^N}}^a - b_{\frac{rt}{2^N}}^a \right). \qquad (15.22)$$

Provided that A_t and B_t satisfy certain analytic conditions these integrals are themselves adapted processes on super Wiener space. Applications of the Itô integral depend heavily on the fact that

$$\int d\mu_b(b_t - b_s) = 0 \qquad (15.23)$$

that is, increments have zero expectation, and also that if $t \geq s$ then the increment $b_t - b_s$ is independent of any adapted process at time less than or equal to s while

$$\int d\mu_b(b_t - b_s)(b_t - b_s) = \tfrac{1}{2}(s - t). \qquad (15.24)$$

These results follow directly from the explicit form of the finite distributions.

The key theorem which underpins the application of stochastic calculus to diffusion equations is the following Itô formula.

Theorem 15.4.2 *Suppose that $Z_t^i, i = 1, \dots, p+q$ are stochastic processes on (m, n)-dimensional super Wiener space, such that*

$$Z_t^i - Z_0^i = \int_o^t A_s^i \, ds + \int_o^t \sum_{a=1}^m C_{as}^i db_s^a \qquad (15.25)$$

for some adapted processes $A_t^i, i = 1, \dots, p + q$, $C_{as}^i, i = 1, \dots, p + q, a = 1, \dots, m$, and that Z_t^i is even for $1 \leq i \leq p$ and odd for $p + 1 \leq i \leq p + q$. Then if f is a supersmooth function on $\mathbb{R}_S^{p,q}$

$$f(Z_t) - f(Z_0) =_\mu \int_o^t \left(\sum_{i=1}^{p+q} \sum_{a=1}^m C_{as}^i \partial_i^S f(Z_s) db_s^a + \sum_{i=1}^{p+q} A_s^i \partial_i^S f(Z_s) ds \right.$$

$$\left. + \tfrac{1}{2} \sum_{i,j=1}^{p+q} \sum_{a=1}^m C_{as}^i C_{as}^j \partial_j^S \partial_i^S f(Z_s) ds \right) \qquad (15.26)$$

where $=_\mu$ indicates that the related processes have the same expectation at each t.

This theorem is proved by considering small time intervals and using (15.23) and (15.24).

At this stage Brownian paths on flat superspace have been considered. In the next section it will be shown how Brownian paths may be constructed on more general supermanifolds.

15.5 Brownian paths on supermanifolds

In order to obtain well defined Brownian paths on a supermanifold (and indeed on a classical manifold) stochastic differential equations are required. These equations, which must themselves take covariant form, provide a covariant notion of Brownian paths. That this will not be entirely straightforward is clear from the second order term in the Itô formula (15.26). Even on $\mathbb{R}^m$ stochastic differential equations have a role, since they make possible Feynman-Kac formulae for a wide class of diffusion operators.

Before passing to supermanifolds, the role of stochastic differential equations in providing Brownian paths on classical manifolds is reviewed. Suppose that M is a compact k-dimensional manifold and that $V_a, a = 1, \ldots, m$ are vector fields on M. Also suppose that $x^i, i = 1 \ldots, k$ are local coordinates on a neighbourhood of a point p in M and that on this neighbourhood $V_a = \sum_{i=1}^m V_a^i(x)\frac{\partial}{\partial x^i}$. Then the stochastic differential equation

$$x_t^i = x^i(p) + \int_o^t V_a^i(x_s)\mathrm{db}_s^a + \tfrac{1}{2}V_a^j(x_s)\left(\partial_j V_a^i(x_s)\right)\mathrm{d}s \qquad (15.27)$$

is, by (15.26), covariantly defined. (It is useful to express the integrand in this equation as $V_a^i \circ \mathrm{db}_s$, introducing the Stratonovich product $\circ$.) Some rather technical steps allow solutions to this local equation to be patched together so that a globally defined stochastic differential equation, with unique solution, is constructed. Equipped with a solution to this equation, the Itô formula (15.26) allows one to prove the following Feynman-Kac-Itô formula for the Hamiltonian operator

$$H = -\tfrac{1}{2}\sum_{a=1}^m V_a V_a . \qquad (15.28)$$

(A more general form, with potential term, is also possible.)

Theorem 15.5.1 *Let x_t be the solution to the stochastic differential equation (15.27). Also let f be a smooth, compact-support function on M. Then* *if the Hamiltonian H defined in (15.28) is elliptic,*

$$\exp\left(-Ht\right) f(p) = \int \mathrm{d}\mu_b \, f(x_t)\,. \tag{15.29}$$

(This theorem shows that x_t is the appropriate Brownian path when the flat Laplacian $\sum_{a=1}^m -\frac{1}{2}\partial_a\partial_a$ is replaced by the more general operator H.)

Outline of proof Using the purely bosonic version of the Itô formula (15.26),

$$\int \mathrm{d}\mu_b \left[f(x_t) - f(p)\right]$$

$$= \int \mathrm{d}\mu_b \left[\int_o^t \partial_i f(x_s)\mathrm{d}x_s^i + \frac{1}{2}\int_o^t \sum_{a=1}^m V_a^i(x_s)V_a^j(x_s)\partial_i\partial_j f(x_s)\mathrm{d}s\right]$$

$$= \int \mathrm{d}\mu_b \left[\frac{1}{2}\int_o^t \sum_{a=1}^m (V_a^j(x_s)\partial_j V_a^i(x_s)\partial_i f(x_s) + V_a^i(x_s)V_a^j(x_s)\partial_i\partial_j f(x_s))\,\mathrm{d}s\right]$$

$$= \int \mathrm{d}\mu_b \left[\int_o^t Hf(x_s)\right]\,.$$

$$\tag{15.30}$$

Thus, if U_t is the operator defined by

$$U_t f(p) = \int \mathrm{d}\mu_b \left[f(x_t)\right] \tag{15.31}$$

it has been shown that

$$U_t f = \int_o^t U_s(-Hf)\mathrm{d}s \tag{15.32}$$

so that $U_t = \exp\left(-Ht\right)$ as required. ∎

It has thus been demonstrated that stochastic differential equations allow one to investigate the heat kernel of a diffusion operator of the form $H = -\frac{1}{2}\sum_{a=1}^m V_aV_a$ on manifold. This method is valid for elliptic second order operators on a manifold, and also for operators such as the scalar Laplacian on a Riemannian manifold (although in this case the stochastic paths are on the orthogonal frame bundle of the manifold rather than on the manifold itself, as observed in [48, 81]). The techniques described here for purely classical manifolds can be extended to Brownian paths on certain supermanifolds to provide information about geometrical operators; these methods are described in Chapter 17. Hamiltonians whose evolution may be studied by the methods described here include those of supersymmetric systems, and also those which arise in BRST quantization as in Chapter 16.

As in the bosonic case, the extension of path integral methods to provide rigorous functional integrals in quantum field theory is far from complete, although heuristic methods from physics show that functional integrals involving fermions are extremely powerful tools.

Chapter 16

Supermanifolds and BRST quantization

BRST quantization is a cohomological technique for the quantization of a theory with gauge symmetry. In such a theory the true degrees of freedom are those of the fields of the system modulo some group of gauge transformations. This can lead to a space with a complicated structure, making the quantization problematical. The BRST procedure replaces the complicated space with cohomology groups on a simpler space; instead of reducing the number of fields, the number is actually increased, but the extra degrees of freedom introduced, known as ghosts, are anticommuting rather than commuting (and obey canonical anticommutation relations rather than commutation relations). The action of the theory has further terms added to it, which include the ghosts and their conjugate momenta. The introduction of ghost degrees of freedom of course means that the classical theory is now developed on a supermanifold. In many cases the structure of the gauge group and its action on the system will mean that this supermanifold incorporates some of the geometric features which have been developed in this book.

In this chapter the basic ideas behind the BRST construction in the context of a quantum mechanical system which is invariant under the action of a finite-dimensional Lie group is explained, using the Hamiltonian formalism. The extension of these methods to more complicated, and more realistic, situations is also discussed. Section 16.1 describes symplectic reduction, the process by which the true phase space of a symmetric Hamiltonian system is constructed, BRST cohomology is introduced in Section 16.2 and BRST quantization, including the gauge fixing procedure, is described in Section 16.3 while in Section 16.4 a system with topological symmetry is described, and its BRST quantization shown to lead to important operators in classical geometry.

16.1   Symplectic reduction

The classical dynamics of a system can be formulated in two equivalent
ways, Lagrangian and Hamiltonian. In the Hamiltonian approach, which
leads directly to quantization in both the Schrödinger and the Heisen-
berg pictures, the dynamics takes place on a symplectic manifold known
as phase space. On a $2n$-dimensional symplectic manifold Darboux coordinates $x^i, p_i, i = 1, \ldots, n$ (in terms of which the symplectic form is
$\omega = \sum_{i=1}^n \mathrm{d}p_i \wedge \mathrm{d}x^i$) always exist locally, and correspond to position and
momentum for the classical system.

A very common situation is that the symplectic manifold is the cotangent bundle T^*M of a manifold M. In this situation local Darboux coordinates of a point α (in other words a one-form α at some point y in M)
are simply local coordinates x^i of y and the coefficients p_i in the expansion
$\alpha = \sum_{i=1}^n p_i \mathrm{d}x^i$. Canonical quantization is straightforward, with states
taken to be functions on M, the position operators multiplication by x^i
and momentum operators $-i\frac{\partial}{\partial x^i}$ (where Planck's constant $\hbar$ has been set
to unity).

However when the classical system possesses symmetry the true phase
space of the system, obtained after redundant degrees of freedom have been
removed, is a symplectic manifold of lower dimension which is generally
much more complicated. The true phase space is obtained by a process
known as *symplectic reduction* which will now be described, in the simplest
setting in which a finite-dimensional Lie group G of dimension m acts on
the $2n$-dimensional phase space N (where $m \leq n$). This group action must
have various properties, it must be free, symplectic and Hamiltonian. A
symplectic action is an action which preserves the symplectic form; the
notion of Hamiltonian action will now be described.

The group action is said to be Hamiltonian if there exists a *constraint
map* $T : \mathfrak{g} \to F(N)$, $\xi \mapsto T_\xi$ (where $F(N)$ denotes the space of smooth
functions on N) which satisfies the conditions

$$\mathcal{L}_{\underline{\xi}} f = \{T_\xi, f\}$$
$$T_\xi(gy) = T_{\mathrm{Ad}_g \xi}(y) \tag{16.1}$$

for all f in $F(N)$, y in N and g in G [67]. (Here for each ξ in the Lie
algebra $\mathfrak{g}$ of G, $\underline{\xi}$ denotes the corresponding vector field on N; also gy
denotes the image of the point y in N under the left action of g in G
and $\{,\}$ denotes Poisson bracket with respect to the symplectic form on

N.) This is the standard set up for a constrained Hamiltonian system: the constraint functions are the m functions $T_a \cong T_{\xi_a}$ corresponding to a basis $\{\xi_a | a = 1, \ldots, m\}$ of $\mathfrak{g}$, and the constraint submanifold C is the subset of N consisting of points y such that $T_a(y) = 0$ for $a = 1, \ldots, m$. More intrinsically, C is the set $\phi^{-1}(0)$ where $\phi : N \to \mathfrak{g}^*$ is the *moment map*, defined by

$$\langle \phi(y), \xi \rangle = T_\xi(y) \tag{16.2}$$

for all y in N and ξ in $\mathfrak{g}$ (so that it is the transpose of the constraint map). By the properties (16.1) of the constraint map, C is invariant under the action of G; the Marsden-Weinstein reduction theorem [101] states that the quotient manifold C/G is a symplectic manifold with symplectic form ν determined uniquely by the condition $\pi^* \nu = \iota^* \omega$, where ω is the symplectic form on N, $\iota : C \to N$ is inclusion and $\pi : C \to C/G$ is the canonical projection. The symplectic manifold obtained by this two stage reduction process is the reduced phase space of the system and will be denoted $N /\!/ G$. It is the true phase space of the system and is in general a rather complicated space, even when N is simple, and may not admit a polarisation as required in quantization to determine the position-momentum split. The BRST approach which the central topic of this chapter is a cohomological formulation which leads to a straightforward quantization procedure.

A further advantage of the BRST procedure is that it allows a fully rigorous gauge-fixing procedure. Without the BRST approach, gauge-fixing is attempted by seeking a further set of conditions (in addition to the constraints) which pick out exactly one element in each gauge orbit. The well-known Gribov problem [66] means that in general such conditions do not exist. In the BRST approach this issue takes a different form, and is more generally solvable [125].

The situation described above, with a finite-dimensional Lie group G action defined on the phases space N, is in fact too restrictive for many situations. It is often the case that the action of the finite group can only be identified locally, and even this local action may only be an infinitesimal action at the level of the Lie algebra. There will however be an infinite-dimensional Lie group $\tilde{G}$ which acts globally on the phase space. The reduction process for this situation will now be described; it is in fact rather similar to that for the simpler situation described above, and leads to a similar BRST quantization procedure; the supermanifolds involved are rather more interesting.

Suppose that $\tilde{G}$ is a Lie group which acts symplectically on N and that there is an open cover $\{U_\sigma | \sigma \in \Lambda\}$ of N and, for each σ in Λ, a neighbourhood V_σ of the identity of G such that V_σ acts locally on U_σ in the following sense: there is a map $V_\sigma \times U_\sigma \to N, (g, y) \to gy$ such that if g, h and gh are in V_σ and y, hy are in U_σ then $(gh)y = g(hy)$. It is also required that the local G action is free, although the global $\tilde{G}$ action may have fixed points. Also suppose that this local G action is compatible with the $\tilde{G}$ action in that if $\tilde{\eta}$ is an element of $\tilde{\mathfrak{g}}$, the Lie algebra of $\tilde{G}$, then, for each basis $\{\xi_a | a = 1, \ldots, m\}$ of $\mathfrak{g}$ (the Lie algebra of the finite group G) and each $\sigma \in \Lambda$ there exist m functions $q_{\tilde{\eta}\sigma}^a : U_\sigma \to \mathbb{R}, a = 1, \ldots, m$ such that for every f in $F(N)$

$$\tilde{\eta}f\big|_{U_\sigma} = q_{\tilde{\eta}\sigma}^a \, \underline{\xi}_a f\big|_{U_\sigma} . \tag{16.3}$$

(Here again the notation is used that $\tilde{\eta}$ denotes the vector field on N corresponding to the element $\tilde{\eta}$ of $\tilde{\mathfrak{g}}$, and so on.) This group action is said to be Hamiltonian if both the global $\tilde{G}$ action and the local G action have constraint maps, denoted $\tilde{T}$ and T_σ respectively, with

$$\tilde{T}_{\tilde{\eta}}\big|_{U_\sigma} = q_{\tilde{\eta}\sigma}^a \, T_{\sigma a} . \tag{16.4}$$

The number of independent constraints is equal to the dimension of G rather than $\tilde{G}$.

An example of these structures occurs when the phase space N is the cotangent bundle T^*M of an n-dimensional manifold M, $\tilde{G}$ is the diffeomorphism group of M (which acts naturally on T^*M) and G is the n-dimensional translation group $\mathrm{Tr}(n)$. (As a manifold this group is simply $\mathbb{R}^n$.) The open cover $\{U_\sigma | \sigma \in \Lambda\}$ of T^*M is constructed from an open cover $\{W_\sigma | \sigma \in \Lambda\}$ of M by coordinate neighbourhoods, setting $U_\sigma = T^*W_\sigma$. The local action of $G = \mathrm{Tr}(n)$ is defined by $(x^i, p_j) \to (x^i + t^i, p_j)$ where $x^i, i = 1, \ldots, n$ are local coordinates on W_σ, (x^i, p_i) the corresponding local coordinates on T^*W_σ and $t^i, i = 1, \ldots, n$ is a sufficiently small element of $\mathbb{R}^n$. The local constraint maps for the G action are then $T_i = p_i$. The Lie algebra of the diffeomorphism group of M may be identified with the set of vector fields on M. If Y is a vector field on M with local coordinate expression $Y = Y^i \frac{\partial}{\partial x^i}$, then the global constraint map for η is

$$\tilde{T}_Y = Y^i p_i . \tag{16.5}$$

The two stage process leading to the reduced phase space can be carried out as before; in the case where N has dimension $2n$ and the local group

G has dimension m the reduced phase space will have dimension $2(n-m)$. Allowing the larger group $\tilde{G}$, together with a local rather than global action by the finite group G, provides an explanation of the observed ambiguities in both the set of constraints and the algebra they form [71]. As well as the physicist's language of constraints, and the geometric language of symplectic reduction, these ideas can be cast in the language of Poisson Lie algebras; these linkages are beautifully described by Stasheff in [144].

16.2 BRST cohomology

In this section the BRST procedure for the reduced phase space is described. The formulation of BRST cohomology in the canonical setting was first given by Henneaux [70], and then expressed in a more abstract mathematical setting by Kostant and Sternberg [96] and by Stasheff [145, 144]. The idea is to construct a BRST operator Q whose zero degree cohomology theory agrees with the space of smooth functions $F(N/\!\!/G)$ on the reduced phase space, and also to construct a super phase space so that the BRST operator Q is implemented by Poisson bracket. This section closely follows [96].

The BRST operator is constructed in two stages. First, a super derivation

$$\delta : \Lambda^q(\mathfrak{g}) \otimes F(N) \mapsto \Lambda^{q-1}(\mathfrak{g}) \otimes F(N)$$

is defined by its action on generators:

$$\delta(\pi \otimes 1) = 1 \otimes T_\pi, \qquad \delta(1 \otimes f) = 0 \qquad (16.6)$$

where $\pi \in \mathfrak{g}$ and $f \in F(N)$. It follows immediately that $\delta^2 = 0$. Also $\mathrm{Ker}^0\,\delta = F(N)$ while $\mathrm{Im}^0\,\delta = \delta(\mathfrak{g})F(N)$. Now $\delta(\mathfrak{g})F(N)$ is the ideal of $F(N)$ consisting of functions which vanish on the constraint surface C, and the space of smooth functions on C can be identified with $F(N)$ modulo this ideal. Thus $F(C) \cong \mathrm{H}^0(\delta)$, and the first part of the construction of the BRST operator has been achieved. (As observed by McMullan [105], This is the Koszul complex.)

To complete the construction, suppose that K is a $\mathfrak{g}$ module, and define the operator

$$\mathrm{d} : K \to \mathfrak{g}^* \otimes K$$

by setting $\langle \mathrm{d}\,k, \pi \rangle = \pi\,k$ for all π in $\mathfrak{g}$ and k in K. This operator can be extended to become a super derivation

$$\mathrm{d} : \Lambda^p(\mathfrak{g}^*) \otimes K \to \Lambda^{p+1}(\mathfrak{g}^*) \otimes K$$

by defining $\mathrm{d}\eta$ for η in $\mathfrak{g}^*$ to be the exterior derivative of η regarded as a left invariant one form on G. Using the fact that d on $\mathfrak{g}^*$ is the transpose of the bracket on $\mathfrak{g}$, it can be shown that $\mathrm{d}^2 = 0$. Also, it follows from the definition that $\mathrm{Ker}\,^0\,\mathrm{d}$ is equal to the set $K^{\mathfrak{g}}$ of $\mathfrak{g}$ invariants in K while $\mathrm{Im}^0\,\mathrm{d}$ is zero. Thus $\mathrm{H}^0\mathrm{d}$ is equal to $K^{\mathfrak{g}}$.

Choosing $K = \Lambda\mathfrak{g} \otimes F(N)$, with the $\mathfrak{g}$ action on K defined by

$$\xi(\pi_1 \wedge \cdots \wedge \pi_q \otimes f)$$

$$= \sum_{r=1}^{q} \pi_1 \wedge \cdots \wedge \pi_{r-1} \wedge [\xi, \pi_r] \wedge \pi_{r+1} \wedge \cdots \wedge \pi_q \otimes f$$

$$+ \pi_1 \wedge \cdots \wedge \pi_q \otimes \{T_\xi, f\}\,, \tag{16.7}$$

the $\mathfrak{g}$ action commutes with the action of δ on K, so that δ and d commute and d is well defined on the δ cohomology groups of K. Thus $\mathrm{H}^0(\mathrm{H}^0(\Lambda\mathfrak{g}^* \otimes \Lambda\mathfrak{g} \otimes F(N)))$ is well defined and equal to the $\mathfrak{g}$ invariant elements of $F(C)$, and hence to $F(N/\!\!/G)$. (As observed by Stasheff [145], this is the Chevalley-Eilenberg differential for the Lie algebra cohomology of G.)

The properties of the two differentials may be summarised in the diagram

$$\Lambda^p(\mathfrak{g}^*) \otimes \Lambda^q(\mathfrak{g}) \otimes F(N) \xrightarrow{\delta} \Lambda^p(\mathfrak{g}^*) \otimes \Lambda^{q-1}(\mathfrak{g}) \otimes F(N)$$

$$\mathrm{d} \downarrow$$

$$\Lambda^{p+1}(\mathfrak{g}^*) \otimes \Lambda^q(\mathfrak{g}) \otimes F(N)$$

giving a double complex

$$\mathrm{D} : \Lambda(\mathfrak{g}^*) \otimes \Lambda(\mathfrak{g}) \otimes F(N) \to \Lambda(\mathfrak{g}^*) \otimes \Lambda(\mathfrak{g}) \otimes F(N)\,.$$

with $\mathrm{D} = \mathrm{d} + (-1)^p\delta$. Defining the total degree of an element of $\Lambda^p(\mathfrak{g}^*) \otimes \Lambda^q(\mathfrak{g}) \otimes F(N)$ to be $p - q$, leads to D raising degree by one. Under certain technical assumptions [96] $\mathrm{H}^0\mathrm{D}$ is equal to $\mathrm{H}^0(\mathrm{H}^0(\Lambda\mathfrak{g}^* \otimes \Lambda\mathfrak{g} \otimes F(N)))$, so that a complex has been constructed whose zero cohomology is equal to $F(N/\!\!/G)$, in other words to the observables on the true phase space of the system.

These algebraic constructions can easily be rephrased in terms of anti-commuting variables and super mathematics once it is observed that $\Lambda \mathfrak{g}^* \otimes \Lambda \mathfrak{g} \otimes F(N)$ can be identified with the space of H^∞ functions on the supermanifold $\mathcal{N} = \mathrm{S}(N, E)$ where E is the trivial $2m$-dimensional bundle over N. This supermanifold can be given a symplectic structure by defining the symplectic form to be

$$\omega + \mathrm{d}\pi_a \wedge \mathrm{d}\eta^a \tag{16.8}$$

where $\pi_a, \eta^a, a = 1, \ldots, m$ are the odd coordinates corresponding to the fibres of E. This allows D to be realised by taking Poisson bracket with the BRST function

$$Q = \eta^a T_a - \tfrac{1}{2} C^c_{ab} \eta^a \eta^b \pi_c . \tag{16.9}$$

As a result the Poisson brackets with respect to this form close on the zero cohomology of D and correspond to the Poisson brackets on the reduced phase space. If the G action is local, in the manner described in Section 16.2, then the super phase space $\mathcal{N}$ will be built from a non-trivial bundle over M; the operator D, while defined locally, is independent of the choice of local system and well defined globally. The symplectic structure and the BRST function Q are also well-defined globally.

16.3 BRST quantization

Quantization of this system is straightforward, given a quantization on the original unconstrained phase space N. The Hilbert space of states has the structure $\mathcal{H} \otimes \Lambda(\mathbb{R})^m$, where $\mathcal{H}$ is the space of states for quantization on N. A typical element is $f_{a_1 \ldots a_p} \eta^{a^1} \ldots \eta^{a^p}$ where each $f_{a_1 \ldots a_p}$ is in $\mathcal{H}$ and $\eta^a, a = 1, \ldots, m$ are natural coordinates on $\mathbb{R}^{0,m}$.

Observables on the super phase space will be operators of the form

$$A^{b_1 \ldots b_q}_{a_1 \cdots p} \eta^{a^1} \ldots \eta^{a^p} \pi_{b_1} \ldots \pi_{b_q}$$

where each $A^{b_1 \ldots b_q}_{a_1 \ldots a_p}$ is an observable on N. The observables η^a, which are known as *ghosts*, are represented on states as multiplication operators, while the ghost momenta π_b are represented by

$$\pi_b = -i \frac{\partial}{\partial \eta^b} . \tag{16.10}$$

An obvious but important consequence of this scheme is that the quantized BRST operator $\mathcal{Q}$ has square zero. This makes it possible to implement the BRST cohomology at the quantum level, both for observables and for states. Physical observables are defined to be observables in the operator BRST cohomology: a physical observable A is an observable which commutes with $\mathcal{Q}$ modulo an observable which is itself a commutator with $\mathcal{Q}$. Physical states are then defined to be states annihilated by $\mathcal{Q}$ modulo those in the image of $\mathcal{Q}$. A particular example of an observable is the Hamiltonian; classically the Hamiltonian is a G invariant function on the phase space N, and so has zero Poisson brackets with the constraint functions. The Hamiltonian can then be extended by terms involving ghosts and ghost momenta to give a Hamiltonian H which satisfies $\{Q, H\} = 0$. On quantization the Hamiltonian becomes an operator which commutes with the quantum BRST operator $\mathcal{Q}$, and hence a quantum BRST observable.

In order to implement BRST quantization by path integral methods a mechanism known as gauge fixing is required which ensures that the traces calculated by the path integrals are traces over BRST cohomology classes, and thus over the physical states of the model. The gauge fixing mechanism involves adding a term H_g to the Hamiltonian which, while zero in the operator cohomology, performs the analytic function of ensuring that all necessary operators are trace class so that the cancellations which formally ought to occur because of supersymmetry actually do occur. This gauge-fixing term H_g in the Hamiltonian is the super commutator $[\mathcal{Q}, \chi] = \mathcal{Q}\chi + \chi\mathcal{Q}$ of an odd operator χ known as the gauge-fixing fermion with the BRST operator $\mathcal{Q}$ [125].

The first step in establishing this mechanism is to observe that if A is an even observable for the BRST quantum system, so that $[A, \mathcal{Q}] = 0$, then, as first observed by Schwarz, [138], the supertrace of A over all states is equal to the supertrace of A over the BRST cohomology classes of states. If $H^i(\mathcal{Q})$ denotes the cohomology of $\mathcal{Q}$ at ghost number i, this result may be expressed by the equation

$$\text{Str } A = \sum_{i=0}^{n} (-1)^i \text{tr}_{H^i(\mathcal{Q})} A, \qquad (16.11)$$

where Str A denotes the supertrace of A. A formal proof comes from noticing that there is cancellation in the supertrace between eigenvalues of A corresponding to states which are not in the kernel of $\mathcal{Q}$ and eigenvalues of states which are in the image of $\mathcal{Q}$. This follows from the observation that if $Af = \lambda f$ then $AQf = \lambda Qf$. Since f and Qf have opposite parity, the only eigenvalues which survive to contribute to the supertrace are those corresponding to BRST cohomology classes. (This is a classic example of the supersymmetry cancellation mechanism.)

This naive argument may of course break down if the reordering of the infinite sums in the traces is not valid. However, by combining A with the operator $\exp\left(-H_g\,t\right)$, $t > 0$ with $H_g = [Q, \chi]$ constructed from an appropriate gauge fixing fermion χ it is possible to obtain precisely the desired traces over cohomology classes, provided of course that these traces exist. The key property which the gauge-fixing Hamiltonian H_g must possess is that the operator $\exp\left(-H_g\,t\right)$ is trace class for all positive t. With this condition satisfied the alternating sums in (16.11) are all absolutely convergent, and the reordering used to establish this equation is valid.

16.4   A topological example

In this section the topological particle model introduced by Baulieu and Singer [14] is described; its BRST quantization leads to the supersymmetric model constructed by Witten in connection with Morse theory [162]. Various supermanifolds are involved in this construction.

The model is defined by the action

$$S\left(x(.)\right) = \int_0^t x^*\mathrm{d}h\,. \tag{16.12}$$

Here the field x is a smooth map $x : I \to M$ from I (the interval $[0, t]$ of the real line) into a compact n-dimensional Riemannian manifold M, while h is a smooth function $h : M \to \mathbb{R}$. In terms of local coordinates x^i this action may be written

$$S\left(x(.)\right) = \int_0^t \partial_i h(x(t'))\,\dot{x}^i(t')\,\mathrm{d}t'. \tag{16.13}$$

Clearly this action can be expressed more simply as

$$S\left(x(.)\right) = h(x(t)) - h(x(0))\,. \tag{16.14}$$

This form of the action shows that the model is indeed topological in nature, a related point being that the equation of motion for x is trivially satisfied. However the form of the action (16.13) involving positions and velocities is required for the passage from the Lagrangian to the Hamiltonian form.

While Beaulieu and Singer considered the case $h = 0$, the Witten model is obtained by allowing h to be an arbitrary function [126]; an interesting case is when h is a Morse function on M, that is, a function with isolated critical points.

It is evident that the action (16.13) is highly symmetric, depending only on the endpoints of the path. It might thus be naively supposed that the path integral

$$\int_{\text{paths/symmetries}} \mathcal{D}x(.) \exp\left(S(x(.))\right) \qquad (16.15)$$

would be trivial. This is not in fact the case because the 'measure' $\mathcal{D}x$ is not simply some limit of a product measure, but must be derived by careful canonical quantization of the theory, which is carried out below.

The first step in this process is to investigate the classical Hamiltonian dynamics. From the action (16.13) the Lagrangian of the theory is seen to be

$$\mathcal{L}(x, \dot{x}) = \partial_i h(x)\, \dot{x}^i, \qquad (16.16)$$

so that the Euclidean time Legendre transformation to the phase space T^*M (the cotangent bundle of M) gives as momentum conjugate to x^i

$$p_i = i\frac{\delta \mathcal{L}(x, \dot{x})}{\delta \dot{x}^i} = i\partial_i h(x). \qquad (16.17)$$

The symmetries of the system now manifest themselves as n constraints on the phase space T^*M:

$$T_i \equiv -p_i + i\partial_i h(x)(x) = 0, \qquad i = 1, \ldots, n. \qquad (16.18)$$

The Poisson brackets on the phase space T^*M are obtained from the standard symplectic form $\omega = \mathrm{d}p_i \wedge \mathrm{d}x^i$, so that as usual $\left\{x^i, p_j\right\} = \delta^i_j$. Direct calculation shows that

$$\{T_i, T_j\} = 0\,, \qquad (16.19)$$

so that the constraints locally correspond to a trivial, Abelian Lie algebra, or in physicist's language form an abelian first class system. The Hamiltonian of the system is, by the Euclidean time prescription,

$$H = ip_i\dot{x}^i + \mathcal{L}(x, \dot{x}) = 0, \qquad (16.20)$$

so that the Hamiltonian is also first class, that is $\{H, T_i\} = 0, i = 1, \ldots, n$.

As is standard in a topological theory, the constraints are of a number that seems to preclude any interesting dynamics – in this case the system has a $2n$-dimensional phase space with n first class constraints, so that by naive counting one would expect the corresponding reduced phase space to be trivial. In fact the theory does capture some topological information as will emerge below.

The first indication of this comes from considering classical gauge-fixing, which shows that the reduced phase space, while as expected zero-dimensional, corresponds to the critical points of h. Classically, gauge-fixing conditions are sought which pick out one point in each orbit of this group; in this case a natural choice is the set of n conditions $X^i \equiv g^{ij}(-p_j - i\partial_j h) = 0$. (It is only at the quantum level that a fully rigorous argument for the gauge-fixing procedures will be required.) Taken together, the constraints and the gauge-fixing condition are satisfied when $p_i = 0$ and $\partial_i h(x) = 0$, that is, at the critical points of the manifold. Using the BRST quantization to implement the constraints and gauge-fixing at the quantum level, it will emerge that this finite reduced phase space can provide topological information.

Two supermanifolds are required for the BRST procedure, a super configuration space $\mathcal{M}$ and a super phase space $\mathcal{N}$. The (n, n)-dimensional super configuration space $\mathcal{M}$ is the supermanifold $S(M, TM)$ built from the tangent bundle of M; corresponding to a coordinate patch on M it has even local coordinates denoted x^i and odd local coordinates η^i, $i = 1, \ldots n$. The $(2n, 2n)$-dimensional super phase space $\mathcal{N}$ is the cotangent bundle of $\mathcal{M}$, but with a non-standard choice of coordinate; again coordinate patches correspond to those on M; the even local coordinates are denoted x^i and p_i while the odd local coordinates are denoted η^i and π_i, in each case with i running from 1 to n. On overlapping coordinate patches x^i and η^i transforms as on $\mathcal{M}$ while p_i and π_i transform by the following rule:

$$p_i' = \frac{\partial x^j}{\partial x'^i} p_j, \qquad \pi_i' = \frac{\partial x^j}{\partial x'^i} \pi_j \qquad (16.21)$$

The simplest, and natural, choice of symplectic form on this manifold, which makes π_i the conjugate momentum to η_i, is

$$w_s = \mathrm{d}\left(p_i \wedge \mathrm{d}x^i + \pi_j \wedge D\eta^i\right)$$
$$= \mathrm{d}p_i \wedge \mathrm{d}x^i + D\pi_j \wedge D\eta^i - \frac{1}{2} R_{ijk}{}^l \pi_l \eta^k \mathrm{d}x^i \wedge \mathrm{d}x^j, \qquad (16.22)$$

where the Levi-Civita connection corresponding to the Riemannian metric g has been used, with Christoffel symbols $\Gamma_{ij}{}^l$ and curvature tensor components $R_{ijk}{}^\rho$, so that

$$D\eta^i = \mathrm{d}\eta^i + \Gamma_{jk}{}^i \eta^k \mathrm{d}x^j, \quad D\pi_j = \mathrm{d}\pi_j - \Gamma_{ji}{}^k \pi_k \mathrm{d}x^j. \qquad (16.23)$$

The corresponding Poisson brackets are:

$$\{p_j, x^i\} = \delta^i_j, \qquad \{p_i, p_j\} = R_{ijk}{}^l \pi_l \eta^k$$

$$\{p_i, \eta^j\} = \Gamma_{ik}{}^j \eta^k, \qquad \{p_i, \pi_k\} = -\Gamma_{ik}{}^j \pi_j,$$

$$\text{and} \quad \{\pi_j, \eta^i\} = \delta^i_j. \qquad (16.24)$$

the others being zero. To quantize, wave functions are taken to be functions $\psi(x; \eta)$ on the super configuration space $\mathcal{M}$. The observables x^i and η^i are simply represented by multiplication by these variables, while the momenta p_i and π_j are represented as

$$p_i = -iD_i \equiv -i\left(\frac{\partial}{\partial x_i} + \eta^j \Gamma_{ij}{}^k \frac{\partial}{\partial \eta^k}\right) \quad \text{and} \quad \pi_j = -i\frac{\partial}{\partial \eta^i}. \qquad (16.25)$$

Using the method of Section 16.2, the BRST operator $\mathcal{Q}$ is

$$\mathcal{Q} = \eta^i T_i = i\eta^i \left(\frac{\partial}{\partial x^i} + \partial_i h(x)\right). \qquad (16.26)$$

(The symmetry of the Christoffel symmetry removes its contribution to the operator, as in the exterior derivative of forms.) The classical gauge-fixing functions X^i then suggest the form of the gauge-fixing fermion χ to be

$$\chi = \pi_i x^i = ig^{ij}\pi_j(D_j - \partial_j h). \qquad (16.27)$$

States of the system, in other words H^∞ functions on $\mathcal{M}$, can of course naturally be identified with forms on M via the identification

$$a_{i_1 \ldots i_p}(x)\eta^{i_1} \ldots \eta^{i_p} \leftrightarrow a_{i_1 \ldots i_p}(x)\mathrm{d}x^{i_1} \ldots \mathrm{d}x^{i_p}. \qquad (16.28)$$

(This identification is considered further in Section 17.1.) Under this identification

$$\mathcal{Q} = -i\eta^i \left(\frac{\partial}{\partial x_i} - v_i\right) = i(\mathrm{d} + \eta^i \partial_i h(x)) = ie^h \mathrm{d} e^{-h} \quad \text{and}$$

$$\chi = -ig^{ij}\pi_j\left(\frac{\partial}{\partial x_i} + \eta^j \Gamma_{ij}{}^k \frac{\partial}{\partial \eta^k} + \partial_i h(x)\right) = \delta - ig^{ij}\partial_i h(x)\pi_j = e^h \delta e^{-h}$$

$$\qquad (16.29)$$

where d is exterior differentiation and δ is the adjoint operator to d, that is $\delta = {}^*\mathrm{d}{}^*$ where * is the Hodge star operator. This shows that $\mathcal{Q}$ and χ are the

supersymmetry operators used by Witten in his study of supersymmetry
and Morse theory [162]. (The identification of states with forms also leads
to a natural inner product on states.)

These expressions for $\mathcal{Q}$ and χ simplify the calculation of the explicit
expression for the canonical BRST Hamiltonian $H_g = -\frac{i}{2}[\mathcal{Q}, \chi]$, leading to

$$H_g = \frac{1}{2}(\mathrm{d} + \delta)^2 + \frac{1}{2}g^{ij}\frac{\partial h}{\partial x^i}\frac{\partial h}{\partial x^j} + \frac{i}{2}g^{ik}(\eta^j\pi_j - \pi_j\eta^j)\frac{D^2 h}{Dx^k Dx^j} \qquad (16.30)$$

which is (up to a factor $\frac{1}{2}$) the Hamiltonian used by Witten [162]. Witten
also shows that the mapping $\psi \mapsto e^{-h}\psi$ induces an isomorphism of de
Rham cohomology classes of d and $\mathcal{Q} = e^{-h}\mathrm{d}e^h$, and that forms with zero
H eigenvalue give exactly one representative of each $\mathcal{Q}$ cohomology class.
Since additionally H has the same eigenvalues as the Laplacian $\frac{1}{2}(\mathrm{d} + \delta)^2$,
the gauge-fixing fermion χ is a good one [125].

The topological exploitation of this model involves calculation of the
heat kernel (in physical parlance, matrix elements) of this operator. These
can be carried out rigorously using the super stochastic methods developed
in Chapter 15, as will be briefly described in Section 17.4 [126].

This one model described here gives some indication as to how super
mathematics, and in particular supermanifolds, may be used in BRST quantization. Further material on this rich area includes the BV quantization
scheme [9, 10, 52] which uses odd symplectic structures [94, 155, 90].

Recent work by the author [128] develops what can be described as an
equivariant BRST theory for systems whose constraints algebras are open
in a very specific way. One example of this procedure leads to a model first
given by Witten which gives a new approach to equivariant cohomology and
fixed point theorems, and to localisation. This model is discussed briefly in
Section 17.4.

Chapter 17

Supermanifolds and geometry

Ideas and techniques from super mathematics have been used in a number of ways in the study of classical differential geometry and topology. Some of these methods were developed long before the label 'super' was used for aspects of mathematics possessing Z_2 graded commutativity in some sense. As remarked in the introduction, Grassmann algebras and exterior algebras have a long history; the true beginning of super mathematics can be seen as Cartan's representation of a Clifford algebra on a Grassmann algebra using operators (in the notation of more recent super mathematics) of the form $\theta^i + \frac{\partial}{\partial \theta^i}$. One of the reasons for the effectiveness of supermanifolds in classical geometry (as well as in fermionic and supersymmetric physics) is this possibility of representing operators on Clifford modules as differential operators on supermanifolds. This allows analytic techniques developed for supersymmetric quantum systems to be applied to many geometric situations. In the case of supersymmetric quantum mechanics, these techniques can be made mathematically rigorous in a direct way. In quantum field theory models the mathematics may be more heuristic, but the geometric insights provided have been considerable. A key idea is what is now referred to as a *supertrace*, but seems first to have been used by McKean and Singer [104], who gave a formula which expresses the archetypal supersymmetric cancellation mechanism, and leads to a local version of the Atiyah-Singer index theorem.

Many aspects of the use of super mathematics in classical geometry, particularly relating to equivariant localisation, are already well covered in the literature, for instance in the the books of Berline, Getzler and Vergne [23] and Guillemin and Sternberg [68]. A construction of particular importance is the *superconnection* of Quillen [112]. This is extensively described and applied in [23], and also forms a substantial section in the book of

Tuynman [149], so that it would be superfluous to include an account of this work here.

This chapter considers some aspects of classical geometry where it is specifically supermanifolds which are involved. In the Section 17.1 the natural isomorphism which relates the space of differential forms on a manifold M to the space of G^∞ functions on the supermanifold $S(M, T^*M)$ constructed from the cotangent bundle of M is developed. Given a choice of metric on M, this allows the Hodge de Rham operator $d + \delta$ and the Laplace Beltrami operator $(d + \delta)^2$ to be expressed as differential operators on $S(M, T^*M)$. Geometric Brownian paths are constructed which allow the analysis of these operators which is described in later sections. Section 17.2 describes the relationship between anticommuting variables, Clifford algebras and spinors and explains how spinors may be realised as functions defined locally on $S(M, T^*M)$. Section 17.3 uses the path integral methods of Chapter 15 to study index theory on a manifold M while Section 17.4 includes a brief description of their application to other aspects of geometry.

17.1   Supermanifolds and differential forms

The supermanifold $S(M, T^*M)$ obtained from the cotangent bundle of a manifold M by the construction of Section 8.1 plays an important role in many applications of supermanifold theory to classical geometry. Explicitly, if M has dimension m, then $S(M, T^*M)$ has dimension (m, m), and local coordinates $(x_\alpha^1, \ldots, x_\alpha^m; \xi_\alpha^1, \ldots, \xi_\alpha^m)$ which change according to the rule

$$x_\beta^i(x_\alpha; \xi_\alpha) = \phi_\beta^i(x_\alpha) \qquad i = 1, \ldots, m$$

$$\xi_\beta^j(x_\alpha; \xi_\alpha) = \sum_{k=1}^m \frac{\partial x_\beta^j}{\partial x_\alpha^k}(x_\alpha)\, \xi_\alpha^k \qquad j = 1, \ldots, m. \tag{17.1}$$

Supersmooth functions on this supermanifold are then naturally identified with forms on M; in local coordinates this identification may be expressed as

$$\sum_{\mu \in M_m} f_\mu(x)\xi^\mu \leftrightarrow \sum_{\mu \in M_m} f_\mu(x)dx^\mu, \tag{17.2}$$

so that Berezin integration on the supermanifold corresponds to the standard integration of top forms on the manifold and the exterior derivative takes the form $d = \xi^i \frac{\partial}{\partial x^i}$.

In Section 17.3 the index theorem for the Hirzebruch signature complex is proved. This involves the Hodge de Rham operator $d + \delta$ whose expression in local coordinates (using the supermanifold formalism) is

$$d + \delta = \psi^i \left(\frac{\partial}{\partial x^i} - \Gamma_{ij}{}^k \xi^j \frac{\partial}{\partial \xi^k} \right) \tag{17.3}$$

where $\Gamma_{ij}{}^k$ are the Christoffel symbols for the Levi-Civita connection of g and $\psi^i = \theta^i + g^{ij} \frac{\partial}{\partial \theta^j}$. The proof of the theorem will use the Laplace Beltrami operator which is the square of the Hodge de Rham operator, and has the expression (by the Weitzenbock formula)

$$-2 (d + \delta)^2 = B - R_i^j(x) \xi^i \frac{\partial}{\partial \xi^j} - \tfrac{1}{2} R_{ki}{}^{jl}(x) \xi^i \xi^k \frac{\partial}{\partial \xi^j} \frac{\partial}{\partial \xi^k} \tag{17.4}$$

where $R_{ki}{}^{jl}$ are the components of the curvature of g and B is the Bochner Laplacian

$$B = g^{ij} \left(\mathcal{D}_i \mathcal{D}_j - \Gamma_{ij}{}^k \mathcal{D}_k \right) \tag{17.5}$$

with $\mathcal{D}_i = \frac{\partial}{\partial x^i} - \Gamma_{ij}{}^k \xi^j \frac{\partial}{\partial \xi^k}$. (A proof of this result may be found in [38].)

For geometric applications of superspace path integration, the significant supermanifold is $S(O(M), TM)$, the underlying manifold being the orthonormal frame bundle $O(M)$ of a Riemannian manifold (M, g), with TM induced as a bundle over $O(M)$ by the projection of $O(M)$ onto M. There is a natural definition of Brownian motion on this supermanifold, based on the Brownian motion on manifolds defined by Elworthy [47, 48] and by Ikeda and Watanabe [81]. These authors introduced the stochastic differential equations (17.7) and showed how solutions give a Feynman-Kac-Itô formula for the Bochner Laplacian. Extended to supermanifolds, these Brownian paths will lead to a Feynman-Kac-Itô formula for the Laplace-Beltrami operator on forms on M. For the commuting, even components of the paths, following [47, 48, 81] paths are considered in the bundle of orthonormal frames $O(M)$ of the Riemannian manifold (M, g) starting from a point (x^i, e_a^i) in $O(M)$. Here local coordinates $(x^i : i = 1, \ldots, m)$ on a contractible coordinate neighbourhood U are used; also $\{e_a : a = 1, \ldots, m\}$ is an orthonormal basis of the space $\mathcal{D}(U)$ of vector fields on U and the vierbein components which provide coordinates on $O(M)|_U$ are defined by

$$e_a = e_a^i \frac{\partial}{\partial x^i}. \tag{17.6}$$

The paths $x_t^i, e_{a,t}^i$ are required to satisfy the stochastic differential equations

$$x_t^i = x^i + \int_0^t e_{a,s}^i \circ db_s^a \qquad e_{a,t}^i = e_a^i - \int_0^t e_{b,s}^k e_{a,s}^l \Gamma_{kl}{}^i(x_s) \circ db_s^b$$

$$(17.7)$$

where $\circ$ denotes the Stratonovich product introduced following (15.27). This is an example of the a stochastic differential equation of the form (15.27) with

$$V_a = e_a^i \frac{\partial}{\partial x^i} + e_{a,s}^k e_{b,s}^l \Gamma_{kl}{}^i(x_s) \frac{\partial}{\partial e_b^i}, \qquad (17.8)$$

the canonical horizontal vector fields on $O(M)$. A patching construction allows a global solution to be constructed in the standard way. (Note that almost surely $g^{ij}(x_t) = \sum_{a=1}^m e_{a\,t}^i e_{a\,t}^j$.) The importance of these stochastic differential equations stems from the fact that $-\frac{1}{2}V_a V_a$ is closely related to the scalar Laplacian and, with a fermionic addition defined in (17.14), gives a differential operator whose square is the Bochner Laplacian. The odd, anticommuting components of the paths $\xi_t^i, \pi_{i\,t}, i = 1, \ldots, m$ are defined in terms of the flat fermionic paths θ_t, ρ_t defined in Section 15.2 as

$$\xi_t^i = \xi^i + \sum_{a=1}^m e_{a\,t}^i \theta_t^a, \qquad \pi_{i\,t} = \sum_{j=1}^n \sum_{a=1}^m e_{a\,t}^j g_{ij}(x_t)\rho_t^a. \qquad (17.9)$$

These paths will be used in Section 17.3 to study the heat kernel of the Laplace Beltrami operator. The combined super paths $(x_t^i, e_{a\,t}^i; \xi_t^i, \pi_{i\,t})$ transform covariantly under split coordinate transformations on $S(O(M), TM)$.

17.2 Supermanifolds and spinors

The starting point for much of the connection between supermanifolds and classical geometry is the Clifford algebra relation

$$[\psi^a, \psi^b] = \delta^{ab} \qquad (17.10)$$

where $\psi^a = \theta^a + \frac{\partial}{\partial \theta^a}$. This expression lies behind the expression (17.3) for the Hodge de Rham operator $d + \delta$; it also allows the Dirac operator $\mathcal{D}$ to be expressed as differential operator on a supermanifold, using anticommuting variables to obtain spinor representations as will now be described.

Suppose that $n = 2l$ is an even positive integer, and let F denote the space $H^\infty(\mathbb{R}_S^{0,n}) \otimes \mathbb{C}$. An element of F can then be expressed as the set of polynomials in n anticommuting variables $\xi^1, \ldots; \xi^n$ of the form

$$\sum_{\mu \in M_n} f_\mu \xi^\mu$$

where each f_μ is a complex number. Regarded as a complex vector space F has dimension 2^n. The operators $\psi^a, a = 1, \ldots, n$ then define a representation of the Clifford algebra C_n on F. To define the 2^l-dimensional spinor representation of C_n a subspace of F is required.

For $r = 1, \ldots, l = n/2$ define $\rho_\xi^r = \xi^{2r-1} - i\xi^{2r}$ and $\tilde{\rho}_\xi^r = 1 - i\xi^{2r-1}\xi^{2r}$. Let G be the subspace of F consisting of elements of the form

$$\sum_{\nu \in M_l} g_\nu \rho_\xi^\nu \tilde{\rho}_\xi^{\nu^c}$$

where the g_ν are complex numbers and ν^c denotes the (ordered) complement of ν in $1 \ldots l$. Then G is closed under the action of the Clifford algebra generators $\psi^i, i = 1, \ldots, n$ and an explicit construction of the 2^l-dimensional spinor representation S of C_n has been obtained. This spinor representation, which is constructed from n real anticommuting variables, is useful in constructing geometric representations of the Dirac operator from supermanifolds. The standard representation, which uses l complex anticommuting variables, is less well adapted to this task.

Suppose that (M, g) is a Riemannian manifold of dimension $n = 2l$ which is spin, and that $O^*(M)$ denotes the bundle of oriented orthonormal coframes on M, spin(M) denotes the spin bundle corresponding to a choice of spin structure on M and $\mathcal{S}$ the associated spinor bundle corresponding to the representation S on G. Then the representation S constructed above may be used to represent the Dirac operator for M as a differential operator on the supermanifold $\mathcal{SM} = S(M, O^*(M))$.

Let $\{U_\alpha | \alpha \in \Lambda\}$ be a cover of M by trivialisation neighbourhoods of the bundles $O^*(M)$ and spin(M). Then, if $\pi : \mathcal{S} \to M$ is the projection mapping of the associated spinor bundle, there is a diffeomorphism $h : \pi^{-1}(U_\alpha) \to U_\alpha \times G$ such that $h^1 = \pi$ where h^1 is the first component of h. Suppose that $s : M \to \mathcal{S}$ is a cross-section of $\mathcal{S}$; then for each $\alpha \in \Lambda$ there is a local representative $s_\alpha : U_\alpha \to G$ defined by

$$h \circ s(x) = (x, s_\alpha(x)). \tag{17.11}$$

If, for each $\alpha, \beta \in \Lambda$ such that $U_\alpha \cap U_\beta$ is nonempty, $g_{\alpha\beta} : U_\alpha \cap U_\beta \to \text{Spin}(n)$

is the transition function of spin(M), then on $U_\alpha \cap U_\beta$

$$s_\beta(x) = g_{\alpha\beta} s_\alpha(x). \tag{17.12}$$

To realise this cross section by local functions on the supermanifold $\mathcal{SM} = \mathrm{S}(M, \mathrm{O}^*(M))$, standard split local coordinates are used with $(x_\alpha^i; \eta_\alpha^a), i, a = 1, \ldots, n$ with x_α^i corresponding to local coordinates on M and η_α^a corresponding to an ordered orthonormal basis $(e^a : a = 1, \ldots, m)$ of $\Omega(U_\alpha)$. It is then possible to identify $s_\alpha = \sum_{\nu \in M_l} g_\nu \rho_\xi^\nu \tilde{\rho}_\xi^{\nu^c}$ with the local function on $\mathcal{SM}$ defined by $f_\alpha(x_\alpha; \eta_\alpha) = \sum_{\nu \in M_l} g_\nu(x) \rho_{\eta_\alpha}^\nu \tilde{\rho}_{\eta_\alpha}^{\nu^c}$. Defining the Clifford algebra to act on local functions f_α by $\psi_{\eta_\alpha}^a$, the change of local function from patch to patch corresponds to the transition functions of the spin bundle, and the Dirac operator is then realised as

$$\mathcal{D} = \sum_{a=1}^{n} \psi_{\eta_\alpha}^a \left(e_a^m \frac{\partial}{\partial x^m} + \Gamma_{abc} \psi_{\eta_\alpha}^b \psi_{\eta_\alpha}^c \right) \tag{17.13}$$

on these local functions, where the vierbein e_a^m gives the expansion $e_a = e_a^m \frac{\partial}{\partial x^m}$ of the orthonormal basis e_a of $\mathcal{D}(U_\alpha)$ dual to the basis e^a of $\Omega(U_\alpha)$ and Γ_{abc} are the connection coefficients of the Levi-Civita connection in this basis. (These are sometimes referred to as the 'spin connection' in the physics literature.) These local operators can be consistently patched together to give the standard globally-defined Dirac operator on the cross section s.

17.3 Supersymmetric quantum mechanics and the Atiyah Singer Index theorem

Two new proofs of the local Atiyah-Singer index theorem for a variety of complexes were given by Alvarez-Gaumé [1] and by Friedan and Windey [58] using supersymmetric quantum mechanics. These proofs did not have full mathematical rigour, relying to some extent on somewhat heuristic physicists' methods, but they were extremely direct. In this section it will be explained how the theory of path integration on supermanifolds developed in Chapter 15 can make these proofs rigorous in a direct way. Other proofs of index theorems involving super mathematics have been given, for instance as described in the book of Berline, Getzler and Vergne [23]. The aim here is to present a method which remains close to the physical insight of Alvarez-Gaumé and of Friedan and Windey.

In order to establish a Feynman-Kac-Itô formula for the Laplace-Beltrami operator $(\mathrm{d} + \delta)^2$ on M, vector fields W_a, $a = 1, \ldots, m$ are used which correspond to horizontal vector fields on $\mathrm{S}(O(M), TM)$ regarded as a bundle over M with connection Γ the Levi-Civita connection for the metric g. In a local coordinate system (x^i, e_a^i, ξ^i) on $\mathrm{S}(O(M), TM)$ these vector fields take the form

$$W_a = e_a^i \frac{\partial}{\partial x^i} - e_a^j e_b^k \Gamma_{jk}{}^i(x) \frac{\partial}{\partial e_b^i} - e_a^j \xi^k \Gamma_{jk}{}^i(x) \frac{\partial}{\partial \xi^i}. \tag{17.14}$$

The key property of the vector fields W_a is that, when acting on functions on $\mathrm{S}(O(M), TM)$ which are independent of the e_a^i (that is, on functions of the form $f = g \circ \pi$ where π is the canonical projection from $\mathrm{S}(O(M), TM)$ onto M) they are related to the Laplace-Beltrami operator $L = (\mathrm{d} + \delta)^2$ by

$$L = -\tfrac{1}{2} \left(W_a W_a - R_i^j(x) \xi^i \frac{\partial}{\partial \xi^j} - \frac{1}{2} R_{ki}{}^{jl}(x) \xi^i \xi^k \frac{\partial}{\partial \xi^j} \frac{\partial}{\partial \xi^l} \right) \tag{17.15}$$

as may be seen from the Weitzenbock formula (17.4) together with the fact that the Bochner Laplacian B satisfies

$$B = -\tfrac{1}{2} W_a W_a. \tag{17.16}$$

This leads to the following Feynman-Kac-Itô formula:

Theorem 17.3.1 *Let $(x_t^i, e_{a\,t}^i, \xi_t^i)$ be the paths defined by (17.7) and (17.9) and suppose that $h \in G^\infty \mathrm{S}(M, TM)$. Then*

$$\exp(-Lt)h(x; \xi)$$
$$= \mathbb{E}_{\mathcal{S}} \left[e^{-\int_0^t \frac{1}{2} R_i^j(x_s) \xi_s^i \pi_{j\,s} + \frac{1}{4} R_{ki}{}^{jl} \xi_s^i \xi_s^k \pi_{l\,s} \pi_{j\,s}\, \mathrm{d}s} h(x_t; \xi_t) \right] \tag{17.17}$$

where $L = (\mathrm{d} + \delta)^2$ is the Laplace-Beltrami operator acting on supersmooth functions on $\mathrm{S}(M, TM)$.

Outline of proof Using the Itô formula Theorem 15.4.2 and equation (17.15),

$$\mathbb{E}_{\mathcal{S}} \left[h(x_t; \xi_t) - h(x; \xi) \right]$$
$$= \mathbb{E}_{\mathcal{S}} \left[\int_0^t \tfrac{1}{2} \left(W_a W_a - R_i^j(x) \xi^i \frac{\partial}{\partial \xi^j} - \tfrac{1}{2} R_{ki}{}^{jl}(x) \xi^i \xi^k \frac{\partial}{\partial \xi^j} \frac{\partial}{\partial \xi^l} \right) h(x_s, \xi_s) \right] \mathrm{d}s$$
$$= \mathbb{E}_{\mathcal{S}} \left[\int_0^t -Lh(x_s, \xi_s) \mathrm{d}s \right]. \tag{17.18}$$

Hence, if $f(x; \xi, t) = \mathbb{E}_{\mathcal{S}}[h(x_t, \xi_t)]$,

$$f(x; \xi, t) - f(x; \xi, 0) = \int_0^t -Lf(x; \xi, s)\, \mathrm{d}s \qquad (17.19)$$

and the result follows.   ∎

This theorem allows a proof of the local Atiyah-Singer index theorem for the Hirzebruch signature complex. (It was shown by Atiyah, Bott and Patodi [3] using K-theoretic arguments that the full Atiyah-Singer index theorem can be deduced from the result for the twisted Hirzebruch signature complex. The method proof given here can be extended to this case as in [124].) The starting point is the formula of McKean and Singer [104] which expresses the Hirzebruch signature of a compact Riemann manifold (M, g) as the supertrace of heat kernel of the Laplacian $L = (\mathrm{d} + \delta)^2$ in the following way:

$$\text{Index } (\mathrm{d} + \delta) = \text{Str} \left(\exp(-Lt) \right). \qquad (17.20)$$

In the course of proving this theorem it emerges that the right hand side is independent of t. The significance of this formula in the context of supersymmetry was first appreciated by Witten [161]. Using the identification of the space of differential forms on M with the space $\mathcal{F}$ of supersmooth functions on $S(M, T(M))$, the supertrace is defined in terms of the involution τ on the space of supersmooth functions whose action on $F \in \mathcal{F}$ with $F(x; \xi) = \sum_{\mu \in M_m} F_\mu(x) \xi^\mu$ is given by

$$\tau\left(F\right)(x; \xi)$$
$$= \sum_{\mu \in M_m} \left(\int \mathrm{d}^m \rho \left(\det(g_{ij}(x)) \right)^{-\frac{1}{2}} \exp\left(i \rho^i \xi^j g_{ij}(x) \right) F_\mu(x) \rho^\mu \right). \qquad (17.21)$$

This involution is essentially the Hodge dual. The supertrace Str of an operator K on the space of supersmooth functions on $S(M, T(M))$ is then defined (when it exists) by the formula

$$\text{Str } K = \text{Tr}\left(\tau K\right). \qquad (17.22)$$

The Atiyah-Singer index theorem for the Hirzebruch signature complex can now be stated.

Theorem 17.3.2

$$\text{Index}(d + \delta) = \int_M \left\{ \det\left(\frac{i\Omega/2\pi}{\tanh i\Omega/2\pi} \right)^{\frac{1}{2}} \right\}_m \qquad (17.23)$$

where Ω is the curvature 2-form of the Levi-Civita connection on (M,g), and the brackets $\{\}_m$ indicate projection onto the m-form component of the integrand.

Using the McKean and and Singer formula (17.20) shows that that an equivalent statement of the theorem is

$$\text{Str}(\exp(-Lt)) = \int_M \left\{ \det\left(\frac{i\Omega/2\pi}{\tanh i\Omega/2\pi} \right)^{\frac{1}{2}} \right\}_m \qquad (17.24)$$

where the determinant is taken in $\mathfrak{so}(m, \mathbb{R})$, the Lie algebra where Ω takes values. This theorem certainly holds if the following stronger, local theorem holds.

Theorem 17.3.3 *At each point x in M*

$$\lim_{t \to 0} \text{str}(\exp(-Lt))(x, x) \, \text{dvol}$$

$$= \left\{ \det\left(\frac{i\Omega/2\pi}{\tanh i\Omega/2\pi} \right)^{\frac{1}{2}} \right\}_m \Bigg|_x \qquad (17.25)$$

where str *denotes the $(2^m \times 2^m)$ matrix supertrace relating to the space of coefficients in the expansion of functions of m anticommuting variables.*

The effect of the matrix supertrace is that $\text{str}(\exp(-Lt))(x, y)$ is the kernel of an operator on smooth functions on M.

   The proof of this theorem makes use of the Feynman-Kac-Itô formula of Theorem 17.3.1 to analysed $\exp(-Lt)$, and then (following Getzler [60]), employs Duhamel's formula to extract information about the heat kernel and show that in the limit as t tends to zero only the required term survives.

Proof Using (11.21) it can be seen that if K is a differential operator on the space of supersmooth functions on $S(M, TM)$ then (with tr denoting the matrix trace)

$$
\begin{aligned}
&\mathrm{str}\,K(x,y)\\
&= \mathrm{tr}\,(\tau K(x,y))\\
&= \int \mathrm{d}^m \rho \, \mathrm{d}^m \xi \, K(x,y,\rho,-\xi)\,(\det(g_{ij}(x)))^{-\frac{1}{2}} \exp\left(i\rho^i \xi^j g_{ij}(x)\right).
\end{aligned}
$$

(17.26)

Now the theorem is local, and it is shown by Cycon, Froese, Kirsch and Simon in [38] that at any particular point x of M the manifold can be replaced by $\mathbb{R}^m$ and the n-dimensional vector bundle by $\mathbb{R}^m \times \mathbb{C}^n$, and a metric and connection chosen, so that in the limit as t tends to zero calculation using the Hamiltonian on $\mathbb{R}^m$ gives the same matrix supertrace as that on M. This is done by first choosing a neighbourhood W of x which has compact closure and is a coordinate neighbourhood for M. Now let U be a neighbourhood of x with $\overline{U} \subset W$ and let $\phi : W \to \mathbb{R}^m$ be a system of normal coordinates on W about x which satisfy $\det(g_{ij}) = 1$ throughout W. (The existence of such coordinates are established in [38].) (For simplicity points in W are identified by their coordinates; in particular x becomes 0.) The standard Taylor expansion in normal coordinates about 0 gives [3]

$$
g_{ij}(y) = \delta_{ij} - \frac{1}{3} y^k y^l R_{ijkl}(0) + \ldots
$$

$$
\Gamma_{ij}{}^k(y) = \frac{1}{3} y^l \left(R_{lji}{}^k(0) + R_{lij}{}^k(0) \right) + \ldots .
$$

(17.27)

Now let L_0 be the flat Laplacian $\frac{\partial}{\partial x^i}\frac{\partial}{\partial x^i}$ on $\mathbb{R}^m$ and let $K_t(y,y';\xi;\xi')$ and $K_t^0(y,y';\xi,\xi')$ be the heat kernels of L and L_0 respectively. From Duhamel's formula (as quoted by Getzler in [60]) it can be seen that

$$
\begin{aligned}
&K_t(y,y';\xi,\xi') - K_t^0(y,y';\xi,\xi')\\
&= \int_0^t \mathrm{d}s\, e^{-(t-s)L}(L-L_0)K_s^0(y,y';\xi,\xi')
\end{aligned}
$$

(17.28)

where the differential operators L and L_0 act with respect to the unprimed arguments. Now

$$
\begin{aligned}
&K_s^0(y,y';\xi,\xi')\\
&= \int \mathrm{d}^m \rho \, (2\pi s)^{\frac{m}{2}} \exp\left(-\frac{(y-y')^2}{2s} - i\rho(\xi-\xi') \right).
\end{aligned}
$$

(17.29)

From this it may be deduced by direct calculation that $\mathrm{str}K_t^0(0,0) = 0$ and, using (11.20),

$$\mathrm{str}K_t(0,0) =$$

$$\int d^m\xi d^m\xi' \, \exp -i\xi \cdot \xi' \times \int_0^t ds \left[\left(e^{-(t-s)L}(L - L_0)K_s^0(0,0;\xi,\xi') \right) \right].$$
(17.30)

The Feynman-Kac-Itô formula Theorem 17.3.1 can then be used to evaluate this expression. Rather than proceeding directly, a simplified Laplacian L_1 is now constructed on $\mathbb{R}_S^{m,m}$ with heat kernel K_t^1 which has the property

$$\lim_{t \to 0} \mathrm{str}K_t^1(0,0) = \lim_{t \to 0} \mathrm{str}K_t(0,0)$$

so that the required supertrace can be calculated.

The simplified Laplacian is built from the approximate metric and connections of (17.27), working to first order in y, and with Euclidean metric. Writing $R_{ijkl}{}^l$ for $R_{ijkl}{}^l(0)$ it is

$$L_1 = - \left[\frac{1}{2} \frac{\partial}{\partial x^i} \frac{\partial}{\partial x^i} - \frac{1}{4} R_{ij}{}^{kl} \xi^i \xi^j \frac{\partial}{\partial \xi^l} \frac{\partial}{\partial \xi^k} + \frac{1}{3} R_j^k \xi^j \frac{\partial}{\partial \xi^k} \right.$$

$$- \frac{1}{3} \xi^j x^l \left(R_l{}^k{}_j{}^i + R_{lj}^{ki} \right) \frac{\partial}{\partial \xi^i} \frac{\partial}{\partial x^k}$$

$$\left. - \frac{1}{18} \xi^{k'} \xi^{l'} x^k x^l \left(R_{kj'k'}{}^i + R_{kk'j'}{}^i \right) \left(R_l{}^{j'}{}_{l'}{}^j + R_{ll'}{}^{j'j} \right) \frac{\partial}{\partial \xi^i} \frac{\partial}{\partial \xi^j} \right].$$
(17.31)

The Brownian paths which lead to a Feynman-Kac-Itô formula as in Theorem 17.3.1 are

$$x^{1i}_t = b_t^i, \quad \text{and} \quad \xi^{1i}_t = \xi^i + \theta_t^i, .$$
(17.32)

Using these paths in the Feynman-Kac-Itô formula, it can then be shown (the details are in [124]) that

$$\lim_{t \to 0} K_t(0,0) = \lim_{t \to 0} K_t^1(0,0).$$
(17.33)

By evaluating the supertrace of $\exp(-L_1 t)$ (using flat space path integration techniques for both fermionic and bosonic paths)the local form of the Atiyah-Singer index theorem 17.3.3 will be established.

Using Duhamel's formula and, again following Getzler [60], making the change of variable $\phi = \sqrt{2\pi t}\,\xi$ gives

$$K_t^1(0,0)$$

$$= \mathbb{E}_{\mathcal{S}}\left[\int d^m\phi\, d^m\xi (2\pi t)^{\frac{m}{2}} \int_0^t ds (2\pi s)^{-\frac{m}{2}}\right.$$

$$\left. \times \left\{ \exp\left(-L_1(t-s)\right)(L_1 - L_0)G_s\left(0, \frac{\phi}{\sqrt{2\pi t}}, \xi'\right) \times \exp\left(-i\frac{\phi}{\sqrt{2\pi t}}\xi'\right) \right\} \right]$$

$$(17.34)$$

where

$$G_s(x;\xi,\xi') = \exp-\left(\frac{x^2}{2s} + i\xi\cdot\xi'\right). \qquad (17.35)$$

Hence, using the fact that (for small t) $x_t \sim \sqrt{t}$ while the fermionic paths are of order 1 [124] it can be shown that

$$\lim_{t\to 0} K_t^1(0,0) = \mathbb{E}_{\mathcal{S}}\left[\int d^m\phi\, d^m\xi (2\pi t)^{\frac{m}{2}} \int_0^t ds (2\pi s)^{-\frac{m}{2}}\right.$$

$$\left. \times \left\{ (L_2 - L_0)\, G_s(b_{t-s}^i, \theta_{t-s}^i, \frac{\phi^i}{\sqrt{2\pi t}}) \right\} \right] \quad (17.36)$$

where $L_2 = L_{2x} + L_{2\xi}$ with

$$L_{2x} = -\left(\frac{1}{2}\frac{\partial}{\partial x^i}\frac{\partial}{\partial x^i} + \frac{1}{2}x^l\frac{\phi^j\phi^i}{2\pi t}R_{jil}{}^k\frac{\partial}{\partial x^k}\right.$$

$$\left. + \frac{1}{8}x^k x^l\frac{\phi^i\phi^j\phi^{i'}\phi^{j'}}{(2\pi t)^2}R_{i'ikk'}R_{j'jl}{}^{k'}\right)$$

$$\text{and} \quad L_{2\xi} = -\frac{1}{4}R_{ijkl}\frac{\phi^j\phi^i}{2\pi t}\psi^k\psi^l. \qquad (17.37)$$

Now $\exp\left(-L_{2x}t(0,0)\right)$ can be evaluated using the standard $\mathbb{R}^2$ result [141] that if

$$L = -\frac{1}{2}\frac{\partial}{\partial x^i}\frac{\partial}{\partial x^i} + \frac{iB}{2}\left(x^1\frac{\partial}{\partial x^2} - x^2\frac{\partial}{\partial x^1}\right) + \frac{1}{8}B^2\left((x^1)^2 + (x^2)^2\right)$$

then

$$\exp\left(-Lt(0,0)\right) = \frac{B}{4\pi\sinh(\frac{1}{2}Bt)}. \tag{17.38}$$

Skew-diagonalising $\Omega_k{}^l = \frac{1}{2}\phi^i\phi^j R_{ijk}{}^l$ by 2 by 2 blocks $\begin{pmatrix} 0 & \Omega_k \\ -\Omega_k & 0 \end{pmatrix}, k = 1, \ldots, \frac{1}{2}m$ down the leading diagonal shows that

$$\exp\left(-L_{2x}t(0,0)\right) = \prod_{k=1}^{m/2} \frac{i\Omega_k}{2\pi t}\frac{1}{\sinh(i\Omega_k/2\pi)}. \tag{17.39}$$

Also, using flat fermionic paths or direct calculation shows that

$$\exp\left(-L_{2\xi}t(\xi;\xi')\right) = \int \mathrm{d}^m\rho\Big\{\exp\left(-i\rho(\xi-\xi')\right)$$
$$\times \prod_{k=1}^{m/2}\left(\cosh\left(\frac{i\Omega_k}{2\pi}\right) + (\xi^{2k-1} + i\rho^{2k-1})(\xi^{2k} + i\rho^{2k})\sinh\left(\frac{i\Omega_k}{2\pi}\right)\right)\Big\}.$$
$$\tag{17.40}$$

This leads to

$$\mathrm{str}\,\exp\left(-L_2 t(0,0)\right)$$
$$= \int \mathrm{d}^m\phi \prod_{k=1}^{m/2}\left(\frac{i\Omega_k}{2\pi t}\frac{1}{\sinh\left(\frac{i\Omega_k}{2\pi}\right)}\cosh\left(\frac{i\Omega_k}{2\pi}\right)\right) \tag{17.41}$$

so that

$$\mathrm{str}\,\exp\left(-Lt(0,0)\right)\,\mathrm{dvol} = \left\{\det\left(\frac{i\Omega/2\pi}{\tanh(i\Omega/2\pi)}\right)^{\frac{1}{2}}\right\}_m \tag{17.42}$$

as required. ∎

17.4 Further applications of supermanifolds

In this section the the geometrical implications of the BRST quantization of two highly symmetric quantum mechanical theories will be discussed. The first of these models is the topological particle model considered in Section 16.4. The BRST Hamiltonian (16.30) for this model, which is built from a Morse function h on a manifold M, was first constructed by Witten

[162], who showed that the matrix elements of this model between critical points of the function h could be used to model the cohomology of M.

Calculation of these matrix elements can be done rigorously using stochastic calculus on the supermanifold $S(M, TM)$ [126]. The idea is to use paths y_t^i in M which satisfy the stochastic version of the equation for the steepest descent curves of h. This equation is

$$dy_t^i = dx_t^i - g^{ij}\partial_j h(x_t)dt$$
$$y_0 = x, \tag{17.43}$$

where x^i are the Brownian paths on the Riemann manifold M defined in (17.7), leading to the Feynman-Kac-Itô formula

$$\exp -tH_g\psi(x, \eta) = \int d\mu \exp(-(h(y_t) - h(x)))$$
$$\exp\left(\int_0^t \left(\frac{D^2h}{Dx^i Dx^j}(ig^{jk}(y_s)\xi_s^i\pi_{s\,k} - \tfrac{1}{2}g^{ij}(y_s))\right.\right.$$
$$\left.\left. + R_i^j(y_s)\xi_s^i\pi_{j\,s} + \tfrac{1}{2}R_{ij}^{kl}(y_s)\xi_s^i\xi_s^j\pi_{k\,s}\pi_{l\,s}\right)ds\right)\psi(y_t, \xi_t), \tag{17.44}$$

where ξ_t, π_t are the anticommuting Brownian paths of (17.9). The idea of Witten's construction is to scale h by a large factor u and model the de Rham cohomology of M using differential forms (in supermanifold language, functions $\Psi(X)$ on $S(M, TM)$) which are eigenfunctions of H_g with very low eigenvalues. The rescaling of h means that such functions are concentrated at critical points of h. The Feynman-Kac-Itô formula allows estimation of the heat kernel $\exp(-H_g t)(A, X)$ when $\epsilon(A)$ is a critical point a of h and $\epsilon(X)$ is near a critical point of h with index one higher than that of a. Denoting the low eigenvalue function concentrated near the critical point a of h by Ψ_a, the heat kernel estimate shows that

$$d\Psi_a = \sum_{\Gamma_{ab}} (-1)^{\sigma(\Gamma_{ab})} e^{-(h(b)-h(a))}\Psi_b \tag{17.45}$$

where each Γ_{ab} is a steepest descent curve joining the critical point a with index p to a critical point b with index $p + 1$ and the sum is overall such curves. The sign factor $(-1)^{\sigma(\Gamma_{ab})}$ is determined by the orientation of the coordinates at b relative to the curve Γ_{ab}. This calculation, which determines the de Rham cohomology of M, is carried out in Witten's paper by instanton methods, and has also been carried out using more classical analytic methods [24, 38].

A second model considered by Witten in [162] leads to a Hamiltonian which relates to the Cartan model of the equivariant cohomology of a Riemannian manifold M under an isometric circle action. This model is derived in [128] from a model whose action is

$$S(x(.)) = \int_0^t v \; x^* \omega \qquad (17.46)$$

where v is a constant, x is a path in M and ω is the one from dual to the Killing vector X which generates the circle action. The BRST procedure is more complicated here because the symmetry of the action is reducible.

Many further topological quantum models have been constructed, in which the model is not simply quantum mechanical but involves fields on a space time of dimension two or higher. Functional integral calculations have led to an array of very powerful mathematical results; these methods are not rigorous, but their power suggests that it should be possible to develop a rigorous formulation of the functional integrals concerned. Given that BRST symmetry is used in quantization of these models, one might expect such methods to involve anticommuting variables. A review of many of these ideas may be found in [85]. Some interesting developments in rigorous functional integral methods for two dimensional (and higher) theories have been made by Léandre, including work involving anticommuting variables [97].

A further area of active current research where supermanifolds prove useful is deformation quantization. Examples of this may be found for example in Kontsevich's work on deformation quantization of Poisson manifolds [93].

Appendix A

Notation

Body of Grassmann algebra element x	$\epsilon(x)$
Body of supermanifold $\mathcal{M}$	$\mathcal{M}_{[\emptyset]}$
Even part of V	V_0
Derivative with respect to even variable	∂^E
Derivative with respect to odd variable	∂^O
Grassmann algebra	$\mathbb{R}_S$
Grassmann analytic continuation of function f	$\widehat{f}$
Odd part of V	V_1
Point in $\mathbb{R}_S^{m,n}$	$(x;\xi) = (x^1,\dots,x^m;\xi^1,\dots,\xi^n)$
Set of multi-indices	M_n
Super vector space	$\mathbb{V} = \mathbb{V}_0 \oplus \mathbb{V}_1$
Super algebra	$\mathbb{A}$
Super derivations of $\mathbb{A}$	$\mathrm{Der}(\mathbb{A})$
Super Lie algebra	$\mathfrak{g}$
Super Lie module	$\mathfrak{u}$
Super matrix	$\mathcal{M} = \begin{pmatrix} \mathcal{M}_{0,0} & \mathcal{M}_{0,1} \\ \mathcal{M}_{1,0} & \mathcal{M}_{1,1} \end{pmatrix}$
Supermanifold (concrete)	$\mathcal{M}$
Superspace	$\mathbb{R}_S^{m,n}$
Supermanifold (algebro-geometric)	(M, A)
Supermanifold built from manifold M and vector bundle E	$S(M, E)$
Supersmooth function	G^∞

Bibliography

[1] Alvarez-Gaumé, L. (1983). Supersymmetry and the Atiyah-Singer index theorem, *Comm. Math. Phys.* **90**, pp. 161–173.

[2] Arnowitt, R. and Nath, P. (1976). Riemannian geometry in spaces with Grassman coordinates, *General Relativity and Gravitation* **7**, pp. 89–103.

[3] Atiyah, M., Bott, R. and Patodi, V. (1973). On the heat equation and the index theorem, *Inventiones Math* **19**, pp. 279–330.

[4] Bahraini, A. (2005). *Supersymétrie et geométrie complexe*, Ph.D. thesis, Université de Paris VII.

[5] Baranov, A., Manin, Y., Frolov, I. and Schwarz, A. (1987). A superanalog of the Selberg trace formula and multiloop contributions for fermionic strings, *Communications in Mathematical Physics* **111**, pp. 373–392.

[6] Baranov, A. and Schwarz, A. (1986). Multiloop contributions in string theory, *JETP Letters* **42**, p. 419.

[7] Baranov, A. and Schwarz, A. (1987). On the multiloop contribution to the string theory, *International Journal of Modern Physics A* **2**, p. 1773.

[8] Bartocci, C., Bruzzo, U. and Ruipérez, D. H. (1991). *The geometry of supermanifolds* (Kluwer).

[9] Batalin, I. A. and Vilkovisky, G. (1977). Relatavistic S-matrix of dynamical systems with boson and fermion constraints, *Phys. Lett.* **B69**, p. 309.

[10] Batalin, I. A. and Vilkovisky, G. (1981). Gauge algebra and quantization, *Phys. Lett.* **102B**, pp. 27–31.

[11] Batchelor, M. (1979). The structure of supermanifolds, *Transactions of the American Mathematical Society* **253**, pp. 329–338.

[12] Batchelor, M. (1980). Two approaches to supermanifolds, *Transactions of the American Mathematical Society* **258**, pp. 257–270.

[13] Batchelor, M. (1988). In search of the graded manifold of maps between graded manifolds, in P. Bongaarts and R. Martini (eds.), *Complex differential geometry and supermanifolds in strings and fields, proceedings of the seventh Scheveningen Conference, August 1987, Lecture Notes in Physics*, Vol. 311, pp. 62–113.

[14] Baulieu and Singer, I. (1989). The topological sigma model, *Comm. Math. Phys.* **125**, pp. 227–237.

[15] Belavin, A. and Knizhnik, V. (1986). Algebraic geometry and the geometry of quantum strings, *Physics Letters* **168B**, pp. 201–206.

[16] Berezin, F. (1966). *The method of Second Quantization* (Academic Press).

[17] Berezin, F. (1987). *Introduction to Superanalysis* (Reidel).

[18] Berezin, F. and Kac, G. (1970). Lie groups with commuting and anticommuting parameters, *USSR Sbornik* **11**, p. 311.

[19] Berezin, F. and Leïtes, D. (1976). Supermanifolds, *Soviet Maths Doklady* **16**, pp. 1218–1222.

[20] Berezin, F. A. (1979). Differential forms on supermanifolds, *Jadernaja Fiz* **30**, 4, pp. 1168–1174, translated in Soviet Journal of Nuclear Physics **30** 605-609.

[21] Berkovits, N. (1994). The ten-dimensional Green-Schwarz superstring is a twisted Neveu-Schwarz-Ramond string, *Nucl.Phys.* **B420**, pp. 332–338.

[22] Berkovits, N., Bershadsky, M., Hauer, T., Zhukov, S. and Zwiebach, B. (2000). Superstring theory on $AdS_2 \times S^2$ as a coset supermanifold, *Nucl.Phys.* **B567**, pp. 61–86.

[23] Berline, N., Getzler, E. and Vergne, M. (2004). *Heat Kernels and Dirac Operators*, 2nd edn. (Springer).

[24] Bismut, J.-M. (1986). The Witten complex and the degnerate Morse inequalities, *Journal of Differential Geometry* **23**, pp. 207–240.

[25] Bochner, S. (1946). *Annals of Mathematics* **47**, p. 192.

[26] Boyer, P. and Gitler, S. (1984). The theory of G^∞-supermanifolds, *Trans. Am. Math. Soc.* **285**, 1, pp. 241–267.

[27] Brink, L., Vecchia, P. D. and Howe, P. (1976). A locally supersymmetric and reparametrisation invariant action for the spinning string, *Phys. Lett.* **65B**, pp. 471–474.

[28] Buchbinder, I. and Kuzenko, S. M. (1998). *Ideas and Methods of Supersymmetry and Supergravity*, revised edn. (Institute of Physics Publishing).

[29] Cartan, E. (1937). *La théorie des groupes finis et continus et la géométrie différentielle traitées par la méthode du repère mobile*, *Cahiers scientifique*, Vol. 18 (Gauthier-Villars, Paris).

[30] Chern, S.-S. (1979). *Complex Manifolds without Potential Theory* (Springer).

[31] Cohn, J. (1987). N = 2 super Riemann surfaces, *Nucl. Phys.* **B 284**, p. 34964.

[32] Cohn, P. M. (1957). *Lie Groups* (CUP).

[33] Cooper, F., Khare, A. and Sukhatme, U. (2001). *Supersymmetry in Quantum Mechanics* (World Scientific).

[34] Cornwell, J. (1989). *Group Theory in Physics*, Vol. III Supersymmetries and Infinite-Dimensional Algebras (Academic Press).

[35] Crane, L. and Rabin, J. (1985a). Global properties of supermanifolds, *Comm.Math.Phys.* **100**, p. 141.

[36] Crane, L. and Rabin, J. (1985b). How different are the supermanifolds of Rogers and DeWitt? *Comm.Math.Phys.* **102**, p. 123.

[37] Crane, L. and Rabin, J. (1988). Super Riemann surfaces: uniformization and Teichmüller theory, *Communications in Mathematical Physics* **113**, pp. 601–623.

[38] Cycon, H., Froese, R., Kirsch, W. and Simon, B. (1987). *Schrödinger operators* (Springer).

[39] Deligne, P., Etingof, P., Freed, D., Jeffrey, L., Kazhdan, D., Morgan, J., Morrison, D. and Witten, E. (1999). *Quantum Fields and Strings: A Course For Mathematicians*, Vol. 1 (American Mathematical Society).

[40] Deser, S. and Zumino, B. (1976). Consistent supergravity, *Physics Letters* **62B**, p. 335.

[41] DeWitt, B. (1964). Dynamical theory of groups and fields, in B. DeWitt and R. Stora (eds.), *Relativity, Groups and Topology II*.

[42] DeWitt, B. (1967). *Dynamical Theory of Groups and Fields* (Nauka).

[43] DeWitt, B. (1984,1992). *Supermanifolds* (Cambridge University Press).

[44] D'Hoker, E. and Phong, D. (1989). Conformal scalar fields and chiral splitting on super Riemann surfaces, *Comm. Math. Phys.* **125**, pp. 469–513.

[45] Dolgikh, S. N., Rosly, A. and Schwarz, A. (1990). Supermoduli spaces, *Communications in Mathematical Physics* **135**, pp. 91–100.

[46] Dragon, N. (1979). Torsion and curvature in extended supergravity, *Z. Phys.* **C2**, pp. 29–32.

[47] Elworthy, K. (1978). Stochastic dynamical systems and their flow, in A. Friedman and M. Pinsky (eds.), *Stochastic Analysis*, pp. 79–95, Academic Press, New York.

[48] Elworthy, K. (1982). *Stochastic Differential Equations on Manifolds*, London Mathematical Society Lecture Notes in Mathematics (Cambridge University Press).

[49] Faddeev, L. and Popov, V. (1967). Feynman diagrams for Yang Mills fields, *Phys.Lett.* **B25**, 1, pp. 29–30.

[50] Ferrara, S., Freedman, D. and van Nieuwenhuizen, P. (1976). Progress towards a theory of supergravity, *Phys. Rev.* **D13**, p. 3214.

[51] Ferrara, S. and van Nieuwenhuizen, P. (1978). The auxiliary fields of supergravity, *Phys. Lett.* **B74**, p. 334.

[52] Fisch, J., Henneaux, M., Stasheff, J. and Teitelboim, C. (1989). Existence, uniqueness and cohomology of the classical BRST charge with ghosts of ghosts, *Comm. Math. Phys.* **120**, p. 379.

[53] Freed, D. (1999). *Five lectures on supersymmetry* (American Mathematical Society).

[54] Freund, P. (1986). *Introduction to supersymmetry* (CUP).

[55] Freund, P. and Rabin, J. (1988). Supertori are algebraic curves, *Comm. Math. Phys.* **114**, pp. 131–145.

[56] Friedan, D. (1986). Unified string theories, in M. Green and D. Gross (eds.), *Proceedings of the Workshop on Unified String Theories*.

[57] Friedan, D., Martinec, E. J. and Shenker, S. H. (1986). Conformal invariance, supersymmetry and string theory, *Nucl.Phys.* **B271**, p. 93.

[58] Friedan, D. and Windey, P. (1984). Supersymmetric derivation of the Atiyah-Singer index theorem and the chiral anomaly, *Nuclear Physics B* **235**, pp. 395–416.

[59] Galperin, A., Ivanov, E., Ogievetsky, V. and Sokatchev, E. (2001). *Harmonic superspace* (CUP).

[60] Getzler, E. (1986). A short proof of the Atiyah-Singer index theorem, *Topology* **25**, pp. 111–7.

[61] Giddings, S. and Nelson, P. (1988a). The geometry of super Riemann surfaces, *Comm. Math. Phys.* **116**, pp. 607–634.

[62] Giddings, S. and Nelson, P. (1988b). Line bundles on super Riemann surfaces, *Communications in Mathematical Physics* **116**, pp. 607–634.

[63] Gotz, G., Quella, T. and Schomerus, V. (2005). Tensor products of $psl(2|2)$ representations, Hep-th/0506072.

[64] Green, M. and Schwarz, J. (1984). Anomaly cancellations in super-

symmetric d = 10 gauge theory and superstring theory, *Phys. Lett.* **149**, pp. 117–122.

[65] Green, P. (1982). On holomorphic graded manifolds, *Proc. Am. Math. Soc.* **85**, pp. 587–590.

[66] Gribov, V. N. (1978). Quantization of non-Abelian gauge theories, *Nuclear Physics* **B130**, pp. 1–19.

[67] Guillemin, V. and Sternberg, S. (1984). *Symplectic techniques in physics* (Cambridge).

[68] Guillemin, V. and Sternberg, S. (1991). *Supersymmetry and equivariant de Rham theory* (Springer).

[69] Helein, F. (2006). A representation formula for maps on supermanifolds, Math-ph/0603045.

[70] Henneaux, M. (1985). Hamiltonian form of the path integral for theories with a gauge freedom, *Phys. Rep.* **126**, p. 1.

[71] Henneaux, M. and Teitelboim, C. (1992). *Quantization of Gauge Systems* (Princeton University Press).

[72] Hicks, N. J. (1965). *Notes on differential geometry* (Van Nostrand).

[73] Hodgkin, L. (1987). A direct calculation of super Teichmüller space, *Lett. Math. Phys.* **14**, pp. 57–63.

[74] Hodgkin, L. (1989). Problems of fields on super Riemann surfaces, *Journal of Geometry in Physics* **6**, pp. 333–348.

[75] Hodgkin, L. (1995). The complex structure of supermoduli space, *Class. Quantum Grav.* **12**, pp. 2135–2148.

[76] Howe, P. (1979). Super Weyl transformations in two dimensions, *Journal of Physics A* **12**, pp. 393–402.

[77] Howe, P., Raetzel, O. and Sezgin, E. (1998). On brane actions and superembeddings, *J. High Energy Physics* **9808**, p. 11.

[78] Howe, P. and Sezgin, E. (1997). Superbranes, *Phys. Lett.* **B390**, p. 133.

[79] Hoyos, J., Quiros, M., Mittelbrunn, J. L. and Urries, F. J. D. (1984). Generalized supermanifolds, I, II, III, *J. Math. Phys.* **25**, pp. 833–854.

[80] Huffmann, A. (1994). On representations of super coalgebras, *J. Phys. A: Math. Gen.* **27**, pp. 6421–6431.

[81] Ikeda, N. and Watanabe, S. (1981). *Stochastic differential equations and diffusion processes* (North-Holland).

[82] Inoue, A. and Maeda, Y. (1987). Super oscillatory integrals and a path integral for non-relativistic spinning particle, *Proc. Japan Acad. Ser. A* **63**, pp. 1–3.

[83] Inoue, A. and Maeda, Y. (1991). Foundations of calculus on super

euclidean space $R^{m|n}$ based on a Fréchet-Grassmann algebra, *Kodai Math. J.* **14**, pp. 72–112.

[84] Jadczyk, A. and Pilch, K. (1981). Superspace and supersymmetries, *Comm. Math. Phys.* **78**, pp. 373–390.

[85] J.Mourão (2004). Aspects of the connections between path integrals, quantum field theory, topology and geometry, *Publicaciones de la Real Sociedad Matemática Española* **7**, pp. 3–51.

[86] Kac, V. (1977). Classification of simple Lie superalgebras, *Adv. Math.* **26**, p. 8.

[87] Khrennikov (1996). Distributions and partial diffferential equations on superspace, *Journal of mathematical sciences* **79**, pp. 816–870.

[88] Khudaverdian, H. M. (1998). Odd invariant semidensity and divergence-like operators on an odd symplectic superspace, *Comm. Math. Phys.* **198**, pp. 591–606.

[89] Khudaverdian, H. M. (2004). Semidensities on odd symplectic super-manifolds, *Comm. Math. Phys.* **247**, 2, pp. 353–390.

[90] Khudaverdian, H. M. and Voronov, T. (2006). Differential forms and odd symplectic geometry, DG/0606560.

[91] Kleppe, A. F. and Wainwright, C. (2006). Graded Majorana spinors, *J. Phys. A* **39**, 14, pp. 3787–3799.

[92] Kobayashi, S. and Nomizu, K. (1963). *Foundations of Differential Geometry* (Interscience).

[93] Kontsevich, M. (2003). Deformation quantization of Poisson manifolds, *Letters in Mathematical Physics* **66**, 3, pp. 157–216.

[94] Kontsevich, M. A. M., Schwarz, A. and Zaboronsky, O. (1997). The geometry of the master equation and topological quantum field theory, *International Journal of Modern Physics* **A 12**, p. 1405.

[95] Kostant, B. (1977). Graded manifolds, graded Lie theory and prequantization, in K. Bleuler and A. Reetz (eds.), *Differential geometrical methods in mathematical physics : proceedings of the symposium held at the University of Bonn, July 1-4, 1975, Lecture Notes in Mathematics*, Vol. 570 (Springer), pp. 177–306.

[96] Kostant, B. and Sternberg, S. (1987). Symplectic reduction, BRS cohomology, and infinite-dimensional Clifford algebras, *Annals of Physics* **176**, pp. 49–113.

[97] Léandre, R. (2003). Super Brownian motion on a loop group, *Rep. Math. Phys.* **51**, p. 269274, XXXIVth symposium of Math. Phys. of Torun, R. Mrugala edt.

[98] LeBrun, C. and Rothstein, M. (1988). Moduli of super Riemann surfaces, *Comm. Math. Phys.* **117**, pp. 159–176.

[99] Leĭtes, D. (1980). Introduction to the theory of supermanifolds, *Russian Math. Surveys* **35**, pp. 1–64.

[100] Manin, Y. (1988). *Gauge field theory and complex geometry* (Springer-Verlag).

[101] Marsden, J. and Weinstein, A. (1974). Reduction of symplectic manifolds with symmetry, *Rep. Math. Phys.* **5**, pp. 121–130.

[102] Martin, J. (1959a). The Feynman principle for a Fermi system, *Proceedings of the Royal Society* **A251**, pp. 543–549.

[103] Martin, J. (1959b). Generalized classical dynamics, and the 'classical analogue' of a Fermi oscillator, *Proceedings of the Royal Society* **A251**, pp. 536–542.

[104] McKean, H. and Singer, I. (1967). Curvature and eigenvalues of the Laplacian, *J. Diff. Geo.* **1**, pp. 43–69.

[105] McMullan, D. (1984). Constraints and BRS symmetry, Imperial College Preprint TP/83-84/21.

[106] Myung, Y. S. (1989). Correlation function for N=1 superconformal models on the supertorus, *Phys. Rev. D* **40**, 6, p. 19801986.

[107] Nagamachi, S. and Kobayashi, Y. (1987). Usage of infinite-dimensional nuclear algebras in superanalysis, *Letters in Mathematical Physics* **14**, pp. 15–23.

[108] Penkov, I. (1983). $\mathcal{D}$ modules on supermanifolds, *Invent. Math.* **71**, pp. 501–512.

[109] Pestov, V. (1992). An analytic structure emerging in presence of infinitely many odd coordinates, ArXiv:funct-an/9211008.

[110] Pestov, V. (1993). Soul expansion of G^∞ superfunctions, *Journal of Mathematical Physics* **34**, pp. 3316–3323.

[111] Polyakov, A. (1981). Quantum geometry of bosonic strings, *Pysics Letters* **103B**, pp. 207–210.

[112] Quillen, D. (1985). Superconnections and the Chern character, *Topology* **24**, pp. 89–95.

[113] Rabin, J. (1995). Super elliptic curves, *Journ. Geom. Phys.* **15**, pp. 252–280.

[114] Rabin, J. (1996). Old and new fields on super Riemann surfaces, *Class.Quant.Grav.* **13**, pp. 875–880.

[115] Rabin, J. and Topiwala, P. (1988). Superriemann surfaces are algebraic curves, Print-88-0569 (UC,San Diego), Jun 1988. 32pp.

[116] Rittenberg, V. (1978). A guide to Lie superalgebras, in P. Kramer

and A. Rieckers (eds.), *Group Theoretical Methods in Physics, Proc. VI Inte. Conf., Tübingen 1977, Lecture Notes in Physics*, Vol. 79 (Springer Verlag), p. 3.

[117] Roepstorff, G. (1996). *Path Integral Approach to Quantum Physics*, Texts and Monographs in Physics (Springer).

[118] Rogers, A. (1980). A global theory of supermanifolds, *Journal of Mathematical Physics* **21**, pp. 1352–1365.

[119] Rogers, A. (1981). Some examples of compact supermanifolds with non-Abelian fundamental group, *Journal of Mathematical Physics* **22**, pp. 443–444.

[120] Rogers, A. (1981b). Super Lie groups: global topology and local structure, *Journal of Mathematical Physics* **22**, pp. 939–945.

[121] Rogers, A. (1985). Consistent superspace integration, *Journal of Mathematical Physics* **26**, pp. 385–392.

[122] Rogers, A. (1986). Graded manifolds, supermanifolds and infinite-dimensional grassmann algebras, *Comm. Math. Phys.* **105**, pp. 375–384.

[123] Rogers, A. (1987). Fermionic path integration and Grassmann Brownian motion, *Communications in Mathematical Physics* **113**, pp. 353–368.

[124] Rogers, A. (1992). Stochastic calculus in superspace II: differential forms, supermanifolds and the Atiyah-Singer index theorem, *Journal of Physics A* **25**, pp. 6043–6062.

[125] Rogers, A. (2000a). Gauge fixing and BFV quantization, *Class. Quantum Grav.* **17**, pp. 389–397.

[126] Rogers, A. (2000b). The topological particle and Morse theory, *Class. Quantum Grav.* **17**, pp. 3703–3714.

[127] Rogers, A. (2003). Supersymmetry and Brownian motion on super-manifolds, *Infinite Dimensional Analysis, Quantum Probability and Related Topics* **6**, Supplementary Issue 1, pp. 83–102.

[128] Rogers, A. (2006). Equivariant BRST quantization and reducible symmetries, King's College London preprint, 22 pages, submitted for publication.

[129] Rogers, A. and Langer, M. (1994). New fields on super Riemann surfaces, *Class. Quantum Grav.* **11**, pp. 2619–2626, corrig: **12**:2619-2620.

[130] Rosly, A., Schwarz, A. and Voronov, A. (1988). Geometry of super-conformal manifolds, *Communications in Mathematical Physics* **119**, pp. 129–152.

[131] Rosly, A., Schwarz, A. and Voronov, A. (1989). Superconformal geometry and string theory, *Communications in Mathematical Physics* **120**, pp. 437–450.

[132] Rothstein, M. (1985). Deformations of complex supermanifolds, *Proc. Americ. Maths. Soc.* **95**, pp. 255–260.

[133] Rothstein, M. (1987). Integration on noncompact supermanifolds, *Trans. Americ. Maths. Soc.* **299**, pp. 387–396.

[134] Rothstein, M. (1991). The structure of supersymplectic supermanifolds, in C. Bartocci, U. Bruzzo and R. Cianci (eds.), *Differential geometric methods in theoretical physics, proceedings, Rapallo 1990, Lecture Notes in Physics*, Vol. 375 (Springer), pp. 331–343.

[135] Salam, A. and Strathdee, J. (1974). Super-gauge transformations, *Nuclear Physics B* **76**, pp. 477–482.

[136] Scheunert, M. (1970). *The Theory of Lie Superalgebras*, Lecture Notes in Mathematics 716 (Springer).

[137] Schulman, L. S. (1981). *Techniques and Applications of Path Integration* (Dover).

[138] Schwarz, A. (1989). Lefschetz trace formula and BRST, *Mod. Phys. Lett.* **A4**, p. 579.

[139] Schwinger, J. (1969). *Particles and sources* (Gordon and Breach, New York).

[140] Shander, V. (1983). Darboux and Liouville theorems on supermanifolds, *Dock. Akad. Nauk, Bulgary* **36**, p. 309.

[141] Simon, B. (1979). *Functional Integration and Quantum mechanics* (Academic Press).

[142] Sorokin, D., Tkach, V. and Volkov, D. (1989a). Superparticles, twistors and Siegel symmetry, *Mod. Phys. Lett.* **A4**, p. 901.

[143] Sorokin, D., Tkach, V., Volkov, D. and Zheltukhin, A. (1989b). From superparticle Siegel symmetry to the spinning particle proper-time supersymmetry, *Phys. Lett.* **B216**, p. 302.

[144] Stasheff, J. (1997). Homological reduction of constrained Poisson algebras, *J. Diff. Geom.* **45**, pp. 221–240.

[145] Stassheff, J. D. (1988). Constrained Poisson algebras and strong homotopy representations,, *Bull. Amer. Math. Soc.* **19**, pp. 287–290.

[146] Stelle, K. and West, P. (1978). Minimal auxiliary fields for supergravity, *Phys. Lett.* **B74**, pp. 330–332.

[147] Sternberg, S. (1964). *Lectures on Differential Geometry* (Prentice Hall).

[148] Swanson, M. (1992). *Path Integrals and Quantum Processes* (Academic Press).

[149] Tuynman, G. M. (2004). *Supermanifolds and super Lie groups*, no. 570 in Mathematics and its Applications (Springer).

[150] Vladimirov, V. and Volovich, I. (1984a). Supermanifolds I. Differential calculus, *Teor. Mat. Fizika* **59**, pp. 3–27.

[151] Vladimirov, V. and Volovich, I. (1984b). Supermanifolds II. Integral calculus, *Teor. Mat. Fizika* **60**, pp. 169–198.

[152] Volkov, D. and Akulov, V. (1973). Is the neutrino a Goldstone particle? *Physics Letters B* **46**, pp. 109–110.

[153] Voronov, T. (1992). Geometric integration theory on supermanifolds, *Soviet Scientific Reviews C: Mathematical Physics Reviews* **9**, pp. 1–138.

[154] Voronov, T. (2002). Dual forms on supermanifolds and Cartan calculus, *Comm. Math. Phys.* **228**, 1, pp. 1–16.

[155] Voronov, T. and Khudaverdian, H. M. (2003). Geometry of differential operators and odd Laplace operators, *Russ. Math. Surv.* **58**, 1, pp. 197–198.

[156] Voronov, T. and Zorich, A. (1986). Complexes of forms on supermanifolds, *Funkts. Analiz i ego Pril.* **20**, 2, pp. 58–59.

[157] Wess, J. and Bagger, J. (1983). *Supersymmetry and Supergravity* (Princeton).

[158] Wess, J. and Zumino, B. (1974). Supergauge transformations in four dimensions, *Nuclear Physics B* **70**, p. 39.

[159] Wess, J. and Zumino, B. (1977). Superspace formulation of supergravity, *Phys. Lett.* **B66**, pp. 361–364.

[160] West, P. (1990). *Introduction to Supersymmetry and Supergravity*, 2nd edn. (World Scientific).

[161] Witten, E. (1982a). Constraints on supersymmetry breaking, *Nuclear Physics B* **202**, p. 253.

[162] Witten, E. (1982b). Supersymmetry and Morse theory, *Journal of Differential Geometry* **17**, pp. 661–692.

[163] Zirnbauer, M. (1991). Fourier analysis on a hyperbolic supermanifold of constant curvature, *Comm. Math. Phys.* **141**, p. 503.

[164] Zumino, B. (1975). Supersymmetry, in R. Arnowitt and P. Nath (eds.), *Proceedings of conference Gauge Theories and Modern Field Theory held at Northeastern University 1975* (MIT press), p. 255.

Index